CAMBRIDGE LIBRARY COLLECTION

Books of enduring scholarly value

Zoology

Until the nineteenth century, the investigation of natural phenomena, plants and animals was considered either the preserve of elite scholars or a pastime for the leisured upper classes. As increasing academic rigour and systematisation was brought to the study of 'natural history', its subdisciplines were adopted into university curricula, and learned societies (such as the London Zoological Society, founded in 1826) were established to support research in these areas. These developments are reflected in the books reissued in this series, which describe the anatomy and characteristics of animals ranging from invertebrates to polar bears, fish to birds, in habitats from Arctic North America to the tropical forests of Malaysia. By the middle of the nineteenth century, this work and developments in research on fossils had resulted in the formulation of the theory of evolution.

Animal Chemistry

At the age of thirteen, chemistry enthusiast Justus von Liebig (1803–73) witnessed the devastation caused by a summer of crop failure. Three decades later, Liebig had become a leading German chemist based at the University of Giessen and had made significant contributions to agriculture and medicine in addition to his pioneering work in organic chemistry. This 1842 study in animal metabolism includes detailed analysis of the chemical transformation undergone in healthy and diseased organisms. Although Liebig considers that chemical analysis alone is not sufficient to explain physiological processes driven by 'vital forces', he argues that it offers quantitative research methods that are superior to mere observation. Several of his works, including this one, were translated into English by his colleague, Scottish chemist William Gregory (1803–58). Liebig's laboratory-based teaching methods quickly gained popularity among British researchers and contributed to the founding of the Royal College of Chemistry in 1845.

Cambridge University Press has long been a pioneer in the reissuing of out-of-print titles from its own backlist, producing digital reprints of books that are still sought after by scholars and students but could not be reprinted economically using traditional technology. The Cambridge Library Collection extends this activity to a wider range of books which are still of importance to researchers and professionals, either for the source material they contain, or as landmarks in the history of their academic discipline.

Drawing from the world-renowned collections in the Cambridge University Library and other partner libraries, and guided by the advice of experts in each subject area, Cambridge University Press is using state-of-the-art scanning machines in its own Printing House to capture the content of each book selected for inclusion. The files are processed to give a consistently clear, crisp image, and the books finished to the high quality standard for which the Press is recognised around the world. The latest print-on-demand technology ensures that the books will remain available indefinitely, and that orders for single or multiple copies can quickly be supplied.

The Cambridge Library Collection brings back to life books of enduring scholarly value (including out-of-copyright works originally issued by other publishers) across a wide range of disciplines in the humanities and social sciences and in science and technology.

Animal Chemistry

*Or, Organic Chemistry in Its Applications
to Physiology and Pathology*

JUSTUS LIEBIG
TRANSLATED BY WILLIAM GREGORY

CAMBRIDGE
UNIVERSITY PRESS

CAMBRIDGE
UNIVERSITY PRESS

University Printing House, Cambridge, CB2 8BS, United Kingdom

Cambridge University Press is part of the University of Cambridge.
It furthers the University's mission by disseminating knowledge in the pursuit of
education, learning and research at the highest international levels of excellence.

www.cambridge.org
Information on this title: www.cambridge.org/9781108080071

© in this compilation Cambridge University Press 2017

This edition first published 1842
This digitally printed version 2017

ISBN 978-1-108-08007-1 Paperback

ANIMAL CHEMISTRY,

OR

ORGANIC CHEMISTRY

IN ITS APPLICATIONS TO

PHYSIOLOGY AND PATHOLOGY.

BY

JUSTUS LIEBIG, M.D., Ph.D., F.R.S., M.R.I.A.

PROFESSOR OF CHEMISTRY IN THE UNIVERSITY OF GIESSEN.

EDITED FROM THE AUTHOR'S MANUSCRIPT

BY WILLIAM GREGORY, M.D., F.R.S.E., M.R.I.A.

PROFESSOR OF MEDICINE AND CHEMISTRY IN THE UNIVERSITY
AND KING'S COLLEGE, ABERDEEN.

LONDON:
PRINTED FOR TAYLOR AND WALTON,
UPPER GOWER STREET.

———

1842.

THE BRITISH ASSOCIATION

ADVANCEMENT OF SCIENCE.

———————

AT the meeting of the British Association in Glasgow, in 1840, I had the honour to present the first part of a report on the then present state of Organic Chemistry, in which I endeavoured to develope the doctrines of this science in their bearing on Agriculture and Physiology.

It affords me now much gratification to be able to communicate to the meeting of the Association for the present year the second part of my labours; in which I have attempted to trace the application of Organic Chemistry to Animal Physiology and Pathology.

In the present work an extensive series of phenomena have been treated in their chemical relations; and although it would be presumptuous to consider the questions here raised as being definitely resolved, yet those who are familiar with chemistry will perceive that the only method which can lead to their final resolution, namely, the *quantitative* method, has been employed.

The formulæ and equations in the second part, therefore, although they are not to be viewed as ascertained truths, and as furnishing a complete, or the only explanation of the vital processes there treated of, are yet true in this sense: that being deduced from facts by logical induction, they must stand as long as no new facts shall be opposed to them.

When the chemist shews, for example, that the elements of the bile, added to those of the urate of ammonia, correspond exactly to those of blood, he presents to us a fact which is independent of all hypothesis. It remains for the physiologist to determine, by experiment, whether the conclusions drawn by the chemist from such a fact be accurate or erroneous. And whether this question be answered in the affirmative or in the negative, the fact remains, and will some day find its true explanation.

I have now to perform the agreeable duty of expressing my sense of the services rendered to me in the preparation of the English edition by my friend Dr. Gregory. The distinguished station he occupies as a chemist; the regular education which he has received in the various branches of medicine; and his intimate acquaintance with the German language—all these, taken together, are the best securities that the translation is such as to convey the exact sense of the original; securities, such as are not often united in the same individual.

It is my intention to follow this second part with a third, the completion of which, however, cannot be looked for before the lapse of two years. This third part will contain an investigation of the food of man and animals, the analysis of all articles of diet, and the study of the changes which the raw food undergoes in its preparation; as, for example, in fermentation (bread), baking, roasting, boiling, &c. Already, it is true, many analyses have been made for the proposed work; but the number of objects of investigation is exceedingly large, and in order to determine with accuracy the absolute value of seed, or of flour, or of a species of fodder, &c., as food, the ultimate analysis alone is not sufficient; there are required comparative investigations, which present very great difficulties.

Dr. JUSTUS LIEBIG.

Giessen,
3rd June, 1842.

NOTE.

I would beg leave to refer the chemical as well as the physiological reader particularly to the analyses (in Note (27), Appendix) of the animal tissues, which ought to have been referred to on pages 43 and 126, and which at present are only referred to in Note (7). Since the work was printed, moreover, there has been added, at the end of the Appendix, an interesting paper by Keller (see page 325), confirming the very important observation of A. Ure, junior, as to the conversion of benzoic acid into hippuric acid in the human body; a fact which I perceive, by the Philosophical Magazine for June, has also been confirmed by Mr. Garrod, probably at an earlier period than by M. Keller. The reader will perceive that this fact strengthens materially the argument of the Author on the action of remedies.

W. G.

PREFACE.

By the application to Chemistry of the methods which had for centuries been followed by philosophers in ascertaining the causes of natural phenomena in physics—by the observation of weight and measure—Lavoisier laid the foundation of a new science, which, having been cultivated by a host of distinguished men, has, in a singularly short period, reached a high degree of perfection.

It was the investigation and determination of all the conditions which are essential to an observation or an experiment, and the discovery of the true principles of scientific research, that protected chemists from error, and conducted them, by a way equally simple and secure, to discoveries which have shed a brilliant light on those natural phenomena which were previously the most obscure and incomprehensible.

The most useful applications to the arts, to industry, and to all branches of knowledge related to chemistry, sprung from the laws thus established; and this influence was not delayed till chemistry had attained its highest perfection, but came into action with each new observation.

All existing experience and observation in other departments of science reacted, in like manner, on the improvement and developement of chemistry; so that chemistry received from metallurgy and from other industrial arts as much benefit as she had conferred on them. While they simultaneously increased in wealth, they mutually contributed to the developement of each other.

After mineral chemistry had gradually attained its present state of developement, the labours of chemists took a new direction. From the study of the constituent parts of vegetables and animals, new and altered views have arisen; and the present work is an attempt to apply these views to physiology and pathology.

In earlier times the attempt has been made, and often with great success, to apply to the objects of the medical art the views derived from an acquaintance with chemical observations. Indeed, the great physicians, who lived towards the end of the seventeenth century, were the founders of chemistry, and in those days the only philosophers acquainted with it. The phlogistic system was the dawn of a new day; it was the victory of philosophy over the rudest empiricism.

With all its discoveries, modern chemistry has performed but slender services to physiology and pathology; and we cannot be deceived as to the cause of this failure, if we reflect that it was found impossible to trace any sort of relation between the observations made in inorganic chemistry, the knowledge of the characters of the elementary bodies and of such of their compounds as could be formed in the laboratory, on the one hand, and the living body, with the characters of its constituents, on the other.

Physiology took no share in the advancement of chemistry, because for a long period she received from the latter science no assistance in her own developement. This state of matters has been entirely changed within five-and-twenty years. But during this period physiology has also acquired new ways and methods of investigation within her own province; and it is only the exhaustion of these sources of discovery which has enabled us to look forward to a change in the direction of the labours of physiologists. The time for such a change is now at hand; and a perseverance in the methods lately followed in physiology would now, from the want, which must soon be felt, of fresh points of departure for researches, render physiology more extensive, but neither more profound nor more solid.

No one will venture to maintain, that the knowledge of the forms and of the phenomena of motion
in organized beings is either unnecessary or unprofitable. On the contrary, this knowledge must be
considered as altogether indispensable to that of the
vital processes. But it embraces only one class of
the conditions necessary for the acquisition of that
knowledge, and is not of itself sufficient to enable
us to attain it.

The study of the uses and functions of the different organs, and of their mutual connection in the
animal body, was formerly the chief object of physiological researches ; but lately this study has fallen
into the back-ground. The greater part of all the
modern discoveries has served to enrich comparative
anatomy far more than physiology.

These researches have yielded the most valuable
results in relation to the recognition of the dissimilar forms and conditions to be found in the healthy
and in the diseased organism ; but they have
yielded no conclusions calculated to give us a more
profound insight into the essence of the vital processes.

The most exact anatomical knowledge of the
structure of the tissues cannot teach us their uses ;
and from the microscopical examination of the most
minute reticulations of the vessels we can learn no

more as to their functions than we have learned
concerning vision from counting the surfaces on the
eye of the fly. The most beautiful and elevated
problem for the human intellect, the discovery of
the laws of vitality, cannot be resolved, nay, cannot
even be imagined, without an accurate knowledge
of chemical forces ; of those forces which do not act
at sensible distances ; which are manifested in the
same way as those ultimate causes by which the
vital phenomena are determined ; and which are
invariably found active, whenever dissimilar substances come into contact.

Physiology, even in the present day, still endea-
vours, but always after the fashion of the phlogistic
chemists (that is, by the *qualitative* method), to
apply chemical experience to the removal of diseased
conditions ; but with all these countless experiments we are not one step nearer to the causes and
the essence of disease.

Without proposing well-defined questions, experi-
menters have placed blood, urine, and all the constituents of the healthy or diseased frame, in contact
with acids, alkalies, and all sorts of chemical re-
agents ; and have drawn, from observation of the
changes thus produced, conclusions as to their
behaviour in the body.

By pursuing this method, useful remedies or modes of treatment might by accident be discovered ; but a rational physiology cannot be founded on mere re-actions, and the living body cannot be viewed as a chemical laboratory.

In certain diseased conditions, in which the blood acquires a viscid consistence, this state cannot be permanently removed by a chemical action on the fluid circulating in the blood-vessels. The deposit of a sediment from the urine may, perhaps, be prevented by alkalies, while their action has not the remotest tendency to remove the cause of disease. Again, when we observe, in typhus, insoluble salts of ammonia in the fæces, and a change in the globules of the blood similar to that which may be artificially produced by ammonia, we are not, on that account, entitled to consider the presence of ammonia in the body as the cause, but only as the effect of a cause.

Thus medicine, after the fashion of the Aristotelian philosophy, has formed certain conceptions in regard to nutrition and sanguification ; articles of diet have been divided into nutritious and non-nutritious ; but these theories, being founded on observations destitute of the conditions most essential to the drawing of just conclusions, could not be received as expressions of the truth.

How clear are now to us the relations of the different articles of food to the objects which they serve in the body, since organic chemistry has applied to the investigation her *quantitative method* of research!

When a lean goose, weighing 4 lbs., gains, in thirty-six days, during which it has been fed with 24 lbs. of maize, 5 lbs. in weight, and yields $3\frac{1}{2}$ lbs. of pure fat, this fat cannot have been contained in the food, ready formed, because maize does not contain the thousandth part of its weight of fat, or of any substance resembling fat. And when a certain number of bees, the weight of which is exactly known, being fed with pure honey, devoid of wax, yield one part of wax for every twenty parts of honey consumed, without any change being perceptible in their health or in their weight, it is impossible any longer to entertain doubt as to the formation of fat from sugar in the animal body.

We must adopt the method which has thus led to the discovery of the origin of fat, in the investigation of the origin and alteration of the secretions, as well as in the study of all the other phenomena of the animal body. From the moment that we begin to look earnestly and conscientiously for the true answers to our questions, that we take the trouble, by means of weight and measure, to fix our

observations, and express them in the form of
equations, these answers are obtained without difficulty.

However numerous our observations may be, yet,
if they only bear on one side of a question, they
will never enable us to penetrate the essence of a
natural phenomenon in its full significance. If we
are to derive any advantage from them, they must
be directed to a well-defined object; and there
must be an organized connection between them.

Mechanical philosophers and chemists justly
ascribe to their methods of research the greater
part of the success which has attended their labours.
The result of every such investigation, if it bear in
any degree the stamp of perfection, may always be
given in few words; but these few words are eternal truths, to the discovery of which numberless
experiments and questions were essential. The
researches themselves, the laborious experiments
and complicated apparatus, are forgotten as soon as
the truth is ascertained. They were the ladders,
the shafts, the tools, which were indispensable to
enable us to attain to the rich vein of ore; they
were the pillars and air passages which protected
the mine from water and from foul air.

Every chemical or physical investigation, however insignificant, which lays claim to attention,

must in the present day possess this character. From a certain number of observations it must enable us to draw some conclusion, whether it be extended or limited.

The imperfection of the method or system of research adopted by physiologists can alone explain the fact, that for the last fifty years they have established so few new and solid truths in regard to a more profound knowledge of the functions of the most important organs, of the spleen, of the liver, and of the numerous glands of the body; and the limited acquaintance of physiologists with the methods of research employed in chemistry will continue to be the chief impediment to the progress of physiology, as well as a reproach which that science cannot escape.

Before the time of Lavoisier, Scheele, and Priestley, chemistry was not more closely related to physics than she is now to physiology. At the present day chemistry is so fused, as it were, into physics, that it would be a difficult matter to draw the line between them distinctly. The connection between chemistry and physiology is the same, and in another half-century it will be found impossible to separate them.

Our questions and our experiments intersect in numberless curved lines the straight line that leads to truth.   It is the points of intersection that indicate to us the true direction; but, owing to the imperfection of the human intellect, these curve lines must be pursued.   Observers in chemistry and physics have the eye ever fixed on the object which they seek to attain.   One may succeed, for a space, in following the direct line; but all are prepared for circuitous paths.   Never doubting of the ultimate success of their efforts, provided they exhibit constancy and perseverance, their eagerness and courage are only exalted by difficulties.

Detached observations, without connection, are points scattered over the plain, which do not allow us to choose a decided path.   For centuries chemistry presented nothing but these points, and sufficient means were available to fill up the intervals between them.   But permanent discoveries and real progress were only made when chemists ceased to make use of fancy to connect them.

My object in the present work has been to direct attention to the points of intersection of chemistry with physiology, and to point out those parts in which the sciences become, as it were, mixed up together. It contains a collection of problems, such as chemistry at present requires to be resolved; and a number of conclusions drawn according to the rules of that science from such observations as have been made.

These questions and problems will be resolved : and we cannot doubt that we shall have in that case a new physiology and a rational pathology. Our sounding line, indeed, is not long enough to measure the depths of the sea, but is not on that account less valuable to us : if it assist us, in the mean time, to avoid rocks and shoals, its use is sufficiently obvious. In the hands of the physiologist, organic chemistry must become an intellectual instrument, by means of which he will be enabled to trace the causes of phenomena invisible to the bodily sight ; and if among the results which I have developed or indicated in this work, one alone shall admit of a useful application, I shall consider the object for which it was written as fully attained. The path which has led to it will open up other paths ; and this I consider as the most important object to be gained.

JUSTUS LIEBIG.

GIESSEN, *April*, 1842.

CONTENTS.

PART I.

PART II.

THE METAMORPHOSIS OF TISSUES.

PART III.

APPENDIX.

ORGANIC CHEMISTRY

PHYSIOLOGY AND PATHOLOGY.

———◆———

I. In the animal ovum, as well as in the seed of a plant, we recognize a certain remarkable force, the source of growth, or increase in the mass, and of reproduction, or of supply of the matter consumed; a force in a state of rest. By the action of external influences, by impregnation, by the presence of air and moisture, the condition of static equilibrium of this force is disturbed; entering into a state of motion or activity, it exhibits itself in the production of a series of forms, which, although occasionally bounded by right lines, are yet widely distinct from geometrical forms, such as we observe in crystallised minerals. This force is called the *vital force, vis vitæ* or *vitality*.

The increase of mass in a plant is determined by the occurrence of a decomposition which takes place in certain parts of the plant under the influence of light and heat.

In the vital process, as it goes on in vegetables, it is exclusively inorganic matter which undergoes this decomposition; and if, with the most distin-

B

guished mineralogists, we consider atmospherical
air and certain other gases as minerals, it may be
said that the vital process in vegetables accom-
plishes the transformation of mineral substances into
an organism endued with life; that the mineral becomes part of an organ possessing vital force.

The increase of mass in a living plant implies that
certain component parts of its nourishment become
component parts of the plant; and a comparison of
the chemical composition of the plant with that of
its nourishment makes known to us, with positive
certainty, which of the component parts of the latter
have been assimilated, and which have been rejected.

The observations of vegetable physiologists and
the researches of chemists have mutually contributed to establish the fact, that the growth and
developement of vegetables depend on the elimination of oxygen, which is separated from the other
component parts of their nourishment.

In contradistinction to vegetable life, the life of
animals exhibits itself in the continual absorption
of the oxygen of the air, and its combination with
certain component parts of the animal body.

While no part of an organised being can serve as
food to vegetables, until, by the processes of putrefaction and decay, it has assumed the form of
inorganic matter, the animal organism requires,
for its support and developement, highly organised
atoms.   The food of all animals, in all circumstances, consists of parts of organisms.

Animals are distinguished from vegetables by the faculty of locomotion, and, in general, by the possession of senses.

The existence and activity of these distinguishing faculties depend on certain instruments which are never found in vegetables. Comparative anatomy shews, that the phenomena of motion and sensation depend on certain kinds of apparatus, which have no other relation to each other than this, that they meet in a common centre. The substance of the spinal marrow, the nerves, and the brain, is in its composition, and in its chemical characters, essentially distinct from that of which cellular substance, membranes, muscles, and skin are composed.

Every thing in the animal organism, to which the name of *motion* can be applied, proceeds from the nervous apparatus. The phenomena of motion in vegetables, the circulation of the sap, for example, observed in many of the characeæ, and the closing of flowers and leaves, depend on physical and mechanical causes. A plant is destitute of nerves. Heat and light are the remote causes of motion in vegetables; but in animals we recognize in the nervous apparatus a source of power, capable of renewing itself at every moment of their existence.

While the assimilation of food in vegetables, and the whole process of their formation, are dependant on certain external influences which produce motion, the developement of the animal organism is, to a certain extent, independent of these external

influences, just because the animal body can produce within itself that source of motion which is
indispensable to the vital process.

Assimilation, or the process of formation and
growth—in other words, the passage of matter from
a state of motion to that of rest—goes on in the
same way in animals and in vegetables. In both,
the same cause determines the increase of mass.
This constitutes the true vegetative life, which is
carried on without consciousness.

The activity of vegetative life manifests itself,
in vegetables, with the aid of external influences;
in animals, by means of influences produced within
their organism. Digestion, circulation, secretion,
are no doubt under the influence of the nervous
system; but the force which gives to the germ, the
leaf, and the radical fibres of the vegetable the
same wonderful properties, is the same as that
residing in the secreting membranes and glands of
animals, and which enables every animal organ to
perform its own proper function. It is only the
source of motion that differs in the two great classes
of organised beings.

While the organs of the vital motions are never
wanting in the lowest orders of animals, as in the impregnated germ of the ovum, in which they are developed first of all, we find, in the higher orders of animals, peculiar organs of feeling and sensation, of consciousness and of a higher intellectual existence.

Pathology informs us that the true vegetative life

is in no way dependant on the presence of this apparatus; that the process of nutrition proceeds in those parts of the body where the nerves of sensation and voluntary motion are paralysed, exactly in the same way as in other parts where these nerves are in the normal condition; and, on the other hand, that the most energetic volition is incapable of exerting any influence on the contractions of the heart, on the motion of the intestines, or on the processes of secretion.

The higher phenomena of mental existence cannot, in the present state of science, be referred to their proximate, and still less to their ultimate causes. We only know of them, that they exist; we ascribe them to an immaterial agency, and that, in so far as its manifestations are connected with matter, an agency entirely distinct from the vital force, with which it has nothing in common.

It cannot be denied that this peculiar force exercises a certain influence on the activity of vegetative life, just as other immaterial agents, such as Light, Heat, Electricity, and Magnetism do; but this influence is not of a determinative kind, and manifests itself only as an acceleration, a retarding, or a disturbance of the process of vegetative life. In a manner exactly analogous, the vegetative life reacts on the conscious mental existence.

There are thus two forces which are found in activity together; but consciousness and intellect may be absent in animals as they are in living vegetables, without their vitality being otherwise affected than by the want of a peculiar source of increased energy or of disturbance. Except in regard to this, all the vital chemical processes go on precisely in the same way in man and in the lower animals.

The efforts of philosophers, constantly renewed, to penetrate the relations of the soul to animal life, have all along retarded the progress of physiology. In this attempt men left the province of philosophical research for that of fancy ; physiologists, carried away by imagination, were far from being acquainted with the laws of purely animal life. None of them had a clear conception of the process of developement and nutrition, or of the true cause of death. They professed to explain the most obscure psychological phenomena, and yet they were unable to say what fever is, and in what way quinine acts in curing it.

For the purpose of investigating the laws of vital motion in the animal body, only one condition, namely, the knowledge of the apparatus which serves for its production, was ascertained ; but the substance of the organs, the changes which food undergoes in the living body, its transformation into portions of organs, and its re-conversion into lifeless compounds, the share which the atmosphere takes in the processes of vitality ; all these foundations for future conclusions were still wanting.

What has the soul, what have consciousness and

intellect to do with the developement of the human fœtus, or the fœtus in a fowl's egg? not more, surely, than with the developement of the seeds of a plant. Let us first endeavour to refer to their ultimate causes those phenomena of life which are not psychological; and let us beware of drawing conclusions before we have a groundwork. We know exactly the mechanism of the eye; but neither anatomy nor chemistry will ever explain how the rays of light act on consciousness, so as to produce vision. Natural science has fixed limits which cannot be passed; and it must always be borne in mind that, with all our discoveries, we shall never know what light, electricity, and magnetism are in their essence, because, even of those things which are material, the human intellect has only conceptions. We can ascertain, however, the laws which regulate their motion and rest, because these are manifested in phenomena. In like manner, the laws of vitality, and of all that disturbs, promotes, or alters it, may certainly be discovered, although we shall never learn what life is. Thus the discovery of the laws of gravitation and of the planetary motions led to an entirely new conception of the cause of these phenomena. This conception could not have been formed in all its clearness without a knowledge of the phenomena out of which it was evolved; for, considered by itself, gravity, like light to one born blind, is a mere word, devoid of meaning.

The modern science of physiology has left the track of Aristotle. To the eternal advantage of science, and to the benefit of mankind, it no longer invents a *horror vacui*, a *quinta essentia*, in order to furnish credulous hearers with solutions and explanations of phenomena, whose true connection with others, whose ultimate cause is still unknown.

If we assume that all the phenomena exhibited by the organism of plants and animals are to be ascribed to a peculiar cause, different in its manifestations from all other causes which produce motion or change of condition; if, therefore, we regard the vital force as an independent force, then, in the phenomena of organic life, as in all other phenomena ascribed to the action of forces, we have the *statics*, that is, the state of equilibrium determined by a resistance, and the *dynamics*, of the vital force.

All the parts of the animal body are produced from a peculiar fluid, circulating in its organism, by virtue of an influence residing in every cell, in every organ, or part of an organ. Physiology teaches that all parts of the body were originally blood; or that at least they were brought to the growing organs by means of this fluid.

The most ordinary experience further shews, that at each moment of life, in the animal organism, a continued change of matter, more or less accelerated, is going on; that a part of the structure is transformed into unorganised matter, loses its condition of life, and must be again renewed. Physiology has sufficiently decisive grounds for the opinion, that every motion, every manifestation of force, is the result of a transformation of the structure or of its substance; that every conception, every mental affection, is followed by changes in the chemical nature of the secreted fluids; that every thought, every sensation, is accompanied by a change in the composition of the substance of the brain.

In order to keep up the phenomena of life in animals, certain matters are required, parts of organisms, which we call nourishment. In consequence of a series of alterations, they serve either for the increase of the mass (*nutrition*), or for the supply of the matter consumed (*reproduction*), or, finally, for the production of force.

II. If the first condition of animal life be the assimilation of what is commonly called nourishment, the second is a continual absorption of oxygen from the atmosphere.

Viewed as an object of scientific research, animal life exhibits itself in a series of phenomena, the connection and recurrence of which are determined by the changes which the food and the oxygen absorbed from the atmosphere undergo in the organism under the influence of the vital force.

All vital activity arises from the mutual action of the oxygen of the atmosphere and the elements of the food.

In the processes of nutrition and reproduction, we perceive the passage of matter from the state of motion to that of rest (static equilibrium); under the influence of the nervous system, this matter enters again into a state of motion. The ultimate causes of these different conditions of the vital force are chemical forces.

The cause of the state of rest is a resistance, determined by a force of attraction (combination), which acts between the smallest particles of matter, and is manifested only when these are in actual contact, or at infinitely small distances.

To this peculiar kind of attraction we may of course apply different names; but the chemist calls it *affinity*.

The cause of the state of motion is to be found in a series of changes which the food undergoes in the organism, and these are the results of processes of decomposition, to which either the food itself, or the structures formed from it, or parts of organs, are subjected.

The distinguishing character of vegetable life is a continued passage of matter from the state of motion to that of static equilibrium. While a plant lives, we cannot perceive any cessation in its growth; no part of an organ in the plant diminishes in size. If decomposition occur, it is the result of assimilation. A plant produces within itself no cause of motion; no part of its structure, from any influence residing in its organism, loses its state of vitality, and is converted into unorganised, amorphous compounds; in a word, no waste occurs in vegetables. Waste, in the animal body, is a change in the state or in the composition of some of its parts, and consequently is the result of chemical actions.

The influence of poisons and of remedial agents on the living animal body evidently shews that the chemical decompositions and combinations in the body, which manifest themselves in the phenomena of vitality, may be increased in intensity by chemical forces of analogous character, and retarded or put an end to by those of opposite character; and that we are enabled to exercise an influence on every part of an organ by means of substances possessing a well-defined chemical action.

As, in the closed galvanic circuit, in consequence of certain changes which an inorganic body, a metal, undergoes when placed in contact with an acid, a certain something becomes cognizable by our senses, which we call a current of electricity; so, in the animal body, in consequence of transformations and changes undergone by matter previously constituting a part of the organism, certain phenomena of motion and activity are perceived, and these we call life, or vitality.

The electrical current manifests itself in certain phenomena of attraction and repulsion, which it excites in other bodies naturally motionless, and by the phenomena of the formation and decomposition of chemical compounds, which occur everywhere, when the resistance is not sufficient to arrest the current.

It is from this point of view, and from no other, that chemistry ought to contemplate the phenomena of life. Wonders surround us on every side. The formation of a crystal, of an octahedron, is not less incomprehensible than the production of a leaf or of a muscular fibre; and the production of vermilion from mercury and sulphur is as much an enigma as the formation of an eye from the substance of the blood.

The first conditions of animal life are nutritious matters and oxygen, introduced into the system.

At every moment of his life man is taking oxygen into his system, by means of the organs of respiration; no pause is observable while life continues.

The observations of physiologists have shewn that the body of an adult man, supplied with sufficient food, has neither increased nor diminished in weight at the end of twenty-four hours; yet the quantity of oxygen taken into the system during this period is very considerable.

According to the experiments of Lavoisier, an adult man takes into his system, from the atmosphere, in one year, 746 lbs., according to Menzies, 837 lbs. of oxygen; yet we find his weight, at the beginning and end of the year, either quite the same, or differing, one way or the other, by at most a few pounds. (1)*

What, it may be asked, has become of the enor-

* The Numbers refer to the Appendix.

mous weight of oxygen thus introduced, in the course of a year into the human system ?

This question may be answered satisfactorily : no part of this oxygen remains in the system ; but it is given out again in the form of a compound of carbon or of hydrogen.

The carbon and hydrogen of certain parts of the body have entered into combination with the oxygen introduced through the lungs and through the skin, and have been given out in the forms of carbonic acid gas and the vapour of water.

At every moment, with every expiration, certain quantities of its elements separate from the animal organism, after having entered into combination, within the body, with the oxygen of the atmosphere.

If we assume, with Lavoisier and Séguin, in order to obtain a foundation for our calculation, that an adult man receives into his system daily $32\frac{1}{2}$ oz. (46,037 cubic inches=15,661 grains, French weight) of oxygen, and that the weight of the whole mass of his blood, of which 80 per cent. is water, is 24 lbs.; it then appears, from the known composition of the blood, that, in order to convert the whole of its carbon and hydrogen into carbonic acid and water, 64,103 grains of oxygen are required. This quantity will be taken into the system of an adult in four days five hours. (2)

Whether this oxygen enters into combination with the elements of the blood, or with other parts of the body containing carbon and hydrogen, in either case the conclusion is inevitable, that the body of a man, who daily takes into the system $32\frac{1}{2}$ oz. of oxygen, must receive daily in the shape of nourishment, as much carbon and hydrogen as would suffice to supply 24 lbs. of blood with these elements; it being presupposed that the weight of the body remains unchanged, and that it retains its normal condition as to health.

This supply is furnished in the food.

From the accurate determination of the quantity of carbon daily taken into the system in the food, as well as of that proportion of it which passes out of the body in the fæces and urine, unburned, that is, in some form in which it is not combined with oxygen, it appears that an adult, taking moderate exercise, consumes 13·9 oz. of carbon daily. (3)

These $13\frac{9}{10}$ oz. of carbon escape through the skin and lungs as carbonic acid gas.

For conversion into carbonic acid gas, $13\frac{9}{10}$ oz. of carbon require 37 oz. of oxygen.

According to the analyses of Boussingault (Ann. de Ch. et de Ph. LXXI. p. 136) a horse consumes in twenty-four hours $97\frac{1}{8}$ oz. of carbon, a milch cow $69\frac{9}{10}$ oz. The quantities of carbon here mentioned are those given off from the bodies of these animals in the form of carbonic acid; and it appears from them that the horse consumes, in converting carbon into carbonic acid, 13 lbs. $3\frac{1}{2}$ oz. in twenty-four hours, and the milch cow 11 lbs. $10\frac{3}{4}$ oz. of oxygen in the same time. (4)

Since no part of the oxygen taken into the system is again given off in any other form but that of a compound of carbon or hydrogen; since further, the carbon and hydrogen given off are replaced by carbon. and hydrogen supplied in the food, it is clear that the amount of nourishment required by the animal body must be in a direct ratio to the quantity of oxygen taken into the system.

Two animals, which in equal times take up by means of the lungs and skin unequal quantities of oxygen, consume quantities of the same nourishment which are unequal in the same ratio.

The consumption of oxygen in equal times may be expressed by the number of respirations; it is clear that, in the same individual, the quantity of nourishment required must vary with the force and number of the respirations.

A child, in whom the organs of respiration are naturally very active, requires food oftener than an adult, and bears hunger less easily. A bird, deprived of food, dies on the third day, while a serpent, with its sluggish respiration, can live without food three months and longer.

The number of respirations is smaller in a state of rest than during exercise or work. The quantity of food necessary in both conditions must vary in the same ratio.

An excess of food is incompatible with deficiency in respired oxygen, that is, with deficient exercise;

*

just as violent exercise, which implies an increased supply of food, is incompatible with weak digestive organs.   In either case the health suffers.

But the quantity of oxygen inspired is also affected by the temperature and density of the atmosphere.

The capacity of the chest in an animal is a constant quantity.   At every respiration a quantity of air enters, the volume of which may be considered as uniform ; but its weight, and consequently that of the oxygen it contains, is not constant.   Air is expanded by heat, and contracted by cold, and therefore equal volumes of hot and cold air contain unequal weights of oxygen.   In summer, moreover, atmospherical air contains aqueous vapour, while in winter it is dry; the space occupied by vapour in the warm air is filled up by air itself in winter ; that is, it contains, for the same volume, more oxygen in winter than in summer.

In summer and in winter, at the pole and at the equator, we respire an equal volume of air ; the cold air is warmed during respiration, and acquires the temperature of the body.   To introduce into the lungs a given volume of oxygen, less expenditure of force is necessary in winter than in summer ; and for the same expenditure of force, more oxygen is inspired in winter.

It is obvious, that in an equal number of respirations we consume more oxygen at the level of the sea than on a mountain.   The quantity both of oxygen inspired and of carbonic acid expired, must therefore vary with the height of the barometer.

The oxygen taken into the system is given out again in the same forms, whether in summer or in winter; hence we expire more carbon in cold weather, and when the barometer is high, than we do in warm weather; and we must consume more or less carbon in our food in the same proportion; in Sweden more than in Sicily; and in our more temperate climate a full eighth more in winter than in summer.

Even when we consume equal weights of food in cold and warm countries, infinite wisdom has so arranged, that the articles of food in different climates are most unequal in the proportion of carbon they contain. The fruits on which the natives of the south prefer to feed do not in the fresh state contain more than 12 per cent. of carbon, while the bacon and train oil used by the inhabitants of the arctic regions contain from 66 to 80 per cent. of carbon.

It is no difficult matter, in warm climates, to study moderation in eating, and men can bear hunger for a long time under the equator; but cold and hunger united very soon exhaust the body.

The mutual action between the elements of the food and the oxygen conveyed by the circulation of the blood to every part of the body is THE SOURCE OF ANIMAL HEAT.

C

III. All living creatures, whose existence depends on the absorption of oxygen, possess within themselves a source of heat independent of surrounding objects.

This truth applies to all animals, and extends, besides,.to the germination of seeds, to the flowering of plants, and to the maturation of fruits.

It is only in those parts of the body to which arterial blood, and with it the oxygen absorbed in respiration, is conveyed, that heat is produced. Hair, wool, or feathers do not possess an elevated temperature.

This high temperature of the animal body, or, as it may be called, disengagement of heat, is uniformly and under all circumstances the result of the combination of a combustible substance with oxygen.

In whatever way carbon may combine with oxygen, the act of combination cannot take place without the disengagement of heat. It is a matter of indifference whether the combination take place rapidly or slowly, at a high or at a low temperature; the amount of heat liberated is a constant quantity.

The carbon of the food, which is converted into carbonic acid within the body, must give out exactly as much heat as if it had been directly burnt in the air or in oxygen gas; the only difference is, that the amount of heat produced is diffused over unequal times. In oxygen, the combustion is more rapid, and the heat more intense; in air it is slower, the temperature is not so high, but it continues longer.

It is obvious that the amount of heat liberated must increase or diminish with the quantity of oxygen introduced in equal times by respiration. Those animals which respire frequently, and consequently consume much oxygen, possess a higher temperature than others, which, with a body of equal size to be heated, take into the system less oxygen. The temperature of a child (102°) is higher than that of an adult (99·5°). That of birds (104° to 105·4°) is higher than that of quadrupeds (98·5° to 100·4°) or than that of fishes or amphibia, whose proper temperature is from 2·7° to 3·6° higher than that of the medium in which they live. All animals, strictly speaking, are warm-blooded; but in those only which possess lungs is the temperature of the body quite independent of the surrounding medium. (5)

The most trustworthy observations prove that in all climates, in the temperate zones as well as at the equator or the poles, the temperature of the body in man, and in what are commonly called warm-blooded animals, is invariably the same; yet how different are the circumstances under which they live!

The animal body is a heated mass, which bears the same relation to surrounding objects as any other heated mass. It receives heat when the surrounding objects are hotter, it loses heat when they are colder than itself.

We know that the rapidity of cooling increases with the difference between the temperature of the heated body and that of the surrounding medium ; that is, the colder the surrounding medium the shorter the time required for the cooling of the heated body.

How unequal, then, must be the loss of heat in a man at Palermo, where the external temperature is nearly equal to that of the body, and in the polar regions, where the external temperature is from 70° to 90° lower.

Yet, notwithstanding this extremely unequal loss of heat, experience has shewn that the blood of the inhabitant of the arctic circle has a temperature as high as that of the native of the south, who lives in so different a medium.

This fact, when its true significance is perceived, proves that the heat given off to the surrounding medium is restored within the body with great rapidity. This compensation takes place more rapidly in winter than in summer, at the pole than at the equator.

Now, in different climates the quantity of oxygen introduced into the system of respiration, as has been already shewn, varies according to the temperature of the external air ; the quantity of inspired oxygen increases with the loss of heat by external cooling, and the quantity of carbon or hydrogen necessary to combine with this oxygen must be increased in the same ratio.

It is evident that the supply of the heat lost by cooling is effected by the mutual action of the elements of the food and the inspired oxygen, which combine together. To make use of a familiar, but not on that account a less just illustration, the animal body acts, in this respect, as a furnace, which we supply with fuel. It signifies nothing what intermediate forms food may assume, what changes it may undergo in the body, the last change is uniformly the conversion of its carbon into carbonic acid, and of its hydrogen into water; the unassimilated nitrogen of the food, along with the unburned or unoxidised carbon, is expelled in the urine or in the solid excrements. In order to keep up in the furnace a constant temperature, we must vary the supply of fuel according to the external temperature, that is, according to the supply of oxygen.

In the animal body the food is the fuel; with a proper supply of oxygen we obtain the heat given out during its oxidation or combustion. In winter, when we take exercise in a cold atmosphere, and when consequently the amount of inspired oxygen increases, the necessity for food containing carbon and hydrogen increases in the same ratio; and by gratifying the appetite thus excited, we obtain the most efficient protection against the most piercing cold. A starving man is soon frozen to death; and every one knows that the animals of prey in the arctic regions far exceed in voracity those of the torrid zone.

In cold and temperate climates, the air, which incessantly strives to consume the body, urges man to laborious efforts in order to furnish the means of resistance to its action, while, in hot climates, the necessity of labour to provide food is far less urgent.

Our clothing is merely an equivalent for a certain amount of food. The more warmly we are clothed the less urgent becomes the appetite for food, because the loss of heat by cooling, and consequently the amount of heat to be supplied by the food, is diminished.

If we were to go naked, like certain savage tribes, or if in hunting or fishing we were exposed to the same degree of cold as the Samoyedes, we should be able with ease to consume 10 lbs. of flesh, and perhaps a dozen of tallow candles into the bargain, daily, as warmly clad travellers have related with astonishment of these people. We should then also be able to take the same quantity of brandy or train oil without bad effects, because the carbon and hydrogen of these substances would only suffice to keep up the equilibrium between the external temperature and that of our bodies.

According to the preceding expositions, the quantity of food is regulated by the number of respirations, by the temperature of the air, and by the amount of heat given off to the surrounding medium.

No isolated fact, apparently opposed to this statement, can affect the truth of this natural law.

Without temporary or permanent injury to health, the Neapolitan cannot take more carbon and hydrogen in the shape of food than he expires as carbonic acid and water; and the Esquimaux cannot expire more carbon and hydrogen than he takes into the system as food, unless in a state of disease or of starvation. Let us examine these states a little more closely.

The Englishman in Jamaica sees with regret the disappearance of his appetite, previously a source of frequently recurring enjoyment; and he succeeds, by the use of cayenne pepper and the most powerful stimulants, in enabling himself to take as much food as he was accustomed to eat at home. But the whole of the carbon thus introduced into the system is not consumed; the temperature of the air is too high, and the oppressive heat does not allow him to increase the number of respirations by active exercise, and thus to proportion the waste to the amount of food taken; disease of some kind, therefore, ensues.

On the other hand, England sends her sick, whose diseased digestive organs have in a greater or less degree lost the power of bringing the food into that state in which it is best adapted for oxidation, and therefore furnish less resistance to the oxidising agency of the atmosphere than is required in their native climate, to southern regions, where the amount of inspired oxygen is diminished in so great a proportion; and the result, an improvement in the health, is obvious. The diseased organs of digestion have sufficient power to place the diminished amount of food in equilibrium with the inspired oxygen; in the colder climate, the organs of respiration themselves would have been consumed in furnishing the necessary resistance to the action of the atmospheric oxygen.

In our climate, hepatic diseases, or those arising from excess of carbon, prevail in summer; in winter, pulmonic diseases, or those arising from excess of oxygen, are more frequent.

The cooling of the body, by whatever cause it may be produced, increases the amount of food necessary. The mere exposure to the open air, in a carriage or on the deck of a ship, by increasing radiation and vaporization, increases the loss of heat, and compels us to eat more than usual. The same is true of those who are accustomed to drink large quantities of cold water, which is given off at the temperature of the body, 98·5°. It increases the appetite, and persons of weak constitution find it necessary, by continued exercise, to supply to the system the oxygen required to restore the heat abstracted by the cold water. Loud and long continued speaking, the crying of infants, moist air, all exert a decided and appreciable influence on the amount of food which is taken.

IV. In the foregoing pages, it has been assumed that it is especially carbon and hydrogen which, by combining with oxygen, serve to produce animal heat. In fact, observation proves that the hydrogen of the food plays a not less important part than the carbon.

The whole process of respiration appears most clearly developed, when we consider the state of a man, or other animal, totally deprived of food.

The first effect of starvation is the disappearance of fat, and this fat cannot be traced either in the urine or in the scanty fæces. Its carbon and hydrogen have been given off through the skin and lungs in the form of oxidised products; it is obvious that they have served to support respiration.

In the case of a starving man, $32\frac{1}{2}$ oz. of oxygen enter the system daily, and are given out again in combination with a part of his body. Currie mentions the case of an individual who was unable to swallow, and whose body lost 100 lbs. in weight during a month; and, according to Martell (Trans. Linn. Soc., vol. xi. p. 411), a fat pig, overwhelmed in a slip of earth, lived 160 days without food, and was found to have diminished in weight, in that time, more than 120 lbs. The whole history of hybernating animals, and the well-established facts of the periodical accumulation, in various animals, of fat, which, at other periods, entirely disappears, prove that the oxygen, in the respiratory process, consumes, without exception, all such substances as are capable of entering into combination with it. It combines with whatever is presented to it; and the deficiency of hydrogen is the only reason why carbonic acid is the chief product; for, at the temperature of the body, the affinity of hydrogen for oxygen far surpasses that of carbon for the same element.

We know, in fact, that the graminivora expire a volume of carbonic acid equal to that of the oxygen inspired, while the carnivora, the only class of animals whose food contains fat, inspire more oxygen than is equal in volume to the carbonic acid expired. Exact experiments have shewn, that in many cases only half the volume of oxygen is expired in the form of carbonic acid. These observations cannot be gainsaid, and are far more convincing than those arbitrary and artificially produced phenomena, sometimes called experiments; experiments which, made as too often they are, without regard to the necessary and natural conditions, possess no value, and may be entirely dispensed with; especially when, as in the present case, nature affords the opportunity for observation, and when we make a rational use of that opportunity.

In the progress of starvation, however, it is not only the fat which disappears, but also, by degrees, all such of the solids as are capable of being dissolved. In the wasted bodies of those who have suffered starvation, the muscles are shrunk and unnaturally soft, and have lost their contractility; all those parts of the body which were capable of entering into the state of motion have served to protect the remainder of the frame from the destructive influence of the atmosphere. Towards the end, the particles of the brain begin to undergo the process of oxidation, and delirium, mania, and death close the scene ; that is to say, all resistance to the oxidising power of the atmospheric oxygen ceases, and the chemical process of eremacausis, or decay, commences, in which every part of the body, the bones excepted, enters into combination with oxygen.

The time which is required to cause death by starvation depends on the amount of fat in the body, on the degree of exercise, as in labour or exertion of any kind, on the temperature of the air, and finally, on the presence or absence of water. Through the skin and lungs there escapes a certain quantity of water, and as the presence of water is essential to the continuance of the vital motions, its dissipation hastens death. Cases have occurred, in which a full supply of water being accessible to the sufferer, death has not occurred till after the lapse of twenty days. In one case, life was sustained in this way for the period of sixty days.

In all chronic diseases death is produced by the same cause, namely, the chemical action of the atmosphere. When those substances are wanting, whose function in the organism is to support the process of respiration ; when the diseased organs are incapable of performing their proper function of producing these substances ; when they have lost the power of transforming the food into that shape in which it may, by entering into combination with the oxygen of the air, protect the system from its influence, then, the substance of the organs themselves, the fat of the body, the substance of the muscles, the nerves, and the brain, are unavoidably consumed.*

The true cause of death in these cases is the respiratory process, that is, the action of the atmosphere.

A deficiency of food, and a want of power to convert the food into a part of the organism, are both, equally, a want of resistance ; and this is the negative cause of the cessation of the vital process. The flame is extinguished, because the oil is consumed ; and it is the oxygen of the air which has consumed it.

In many diseases substances are produced which are incapable of assimilation. By the mere deprivation of food, these substances are removed from the body without leaving a trace behind ; their elements have entered into combination with the oxygen of the air.

From the first moment that the function of the lungs or of the skin is interrupted or disturbed, compounds, rich in carbon, appear in the urine, which acquires a brown colour. Over the whole surface of the body oxygen is absorbed, and combines with all the substances which offer no resistance to it. In those parts of the body where the access of

* For an account of what really takes place in this process, I refer to the considerations on the means by which the change of matter is effected in the body of the carnivora, which will be found further on.

oxygen is impeded ; for example, in the arm-pits, or
in the soles of the feet, peculiar compounds are given
out, recognisable by their appearance, or by their
odour. These compounds contain much carbon.

Respiration is the falling weight, the bent spring,
which keeps the clock in motion; the inspirations
and expirations are the strokes of the pendulum
which regulate it. In our ordinary time-pieces, we
know with mathematical accuracy the effect produced on their rate of going, by changes in the
length of the pendulum, or in the external temperature. Few, however, have a clear conception of
the influence of air and temperature on the health
of the human body; and yet the research into the
conditions necessary to keep it in the normal state,
is not more difficult than in the case of a clock.

V. The want of a just conception of force and
effect, and of the connection of natural phenomena,
has led chemists to attribute a part of the heat generated in the animal body to the action of the nervous system. If this view exclude chemical action,
or changes in the arrangement of the elementary
particles, as a condition of nervous agency, it means
nothing else than to derive the presence of motion,
the manifestation of a force, from nothing. But no
force, no power can come of nothing.

No one will seriously deny the share which the
nervous apparatus has in the respiratory process ;
for no change of condition can occur in the body

without the nerves; they are essential to all vital motions. Under their influence, the viscera produce those compounds, which, while they protect the organism from the action of the oxygen of the atmosphere, give rise to animal heat; and when the nerves cease to perform their functions, the whole process of the action of oxygen must assume another form. When the pons Varolii is cut through in the dog, or when a stunning blow is inflicted on the back of the head, the animal continues to respire for some time, often more rapidly than in the normal state; the frequency of the pulse at first rather increases than diminishes, yet the animal cools as rapidly as if sudden death had occurred. Exactly similar observations have been made on the cutting of the spinal cord, and of the par vagum. The respiratory motions continue for a time, but the oxygen does not meet with those substances with which, in the normal state, it would have combined; because the paralyzed viscera will no longer furnish them. The singular idea that the nerves produce animal heat, has obviously arisen from the notion that the inspired oxygen combines with carbon, in the blood itself; in which case the temperature of the body, in the above experiments, certainly, ought not to have sunk. But, as we shall afterwards see, there cannot be a more erroneous conception than this.

As by the division of the pneumogastric nerves the motion of the stomach and the secretion of the gastric juice are arrested, and an immediate check is thus given to the process of digestion, so the paralysis of the organs of vital motion in the abdominal viscera affects the process of respiration. These processes are most intimately connected ; and every disturbance of the nervous system or of the nerves of digestion re-acts visibly on the process of respiration.

The observation has been made, that heat is produced by the contraction of the muscles, just as in a piece of caoutchouc, which, when rapidly drawn out, forcibly contracts again, with disengagement of heat. Some have gone so far as to ascribe a part of the animal heat to the mechanical motions of the body, as if these motions could exist without an expenditure of force consumed in producing them ; how then, we may ask, is this force produced ?

By the combustion of carbon, by the solution of a metal in an acid, by the combination of the two electricities, positive and negative, by the absorption of light, and even by the rubbing of two solid bodies together with a certain degree of rapidity, heat may be produced.

By a number of causes, in appearance entirely distinct, we can thus produce one and the same effect. In combustion and in the production of galvanic electricity we have a change of condition in material particles ; when heat is produced by the absorption of light or by friction, we have the conversion of one kind of motion into another, which affects our senses differently. In all such cases we have a something given, which merely takes another form ; in all we have a force and its effect. By means of the fire which heats the boiler of a steam-engine we can produce every kind of motion, and by a certain amount of motion we can produce fire.

When we rub a piece of sugar briskly on an iron grater, it undergoes, at the surfaces of contact, the same change as if exposed to heat ; and two pieces of ice, when rubbed together, melt at the point of contact.

Let us remember that the most distinguished authorities in physics consider the phenomena of heat as phenomena of motion, because the very conception of the *creation* of matter, even though imponderable, is absolutely irreconcilable with its production by mechanical causes, such as friction or motion.

But, admitting all the influence which electric or magnetic disturbances in the animal body can have on the functions of its organs, still the ultimate cause of all these forces is a change of condition in material particles, which may be expressed by the conversion, within a certain time, of the elements of the food into oxidised products. Such of these elements as do not undergo this process of slow combustion, are given off unburned or incombustible in the excrements.

Now, it is absolutely impossible that a given amount of carbon or hydrogen, whatever different forms they may assume in the progress of the combustion, can produce more heat than if directly burned in atmospheric air or in oxygen gas.

When we kindle a fire under a steam-engine, and employ the power obtained to produce heat by friction, it is impossible that the heat thus obtained can ever be greater than that which was required to heat the boiler; and if we use the galvanic current to produce heat, the amount of heat obtained is never, in any circumstances, greater than we might have by the combustion of the zinc which has been dissolved in the acid.

The contraction of muscles produces heat; but the force necessary for the contraction has manifested itself through the organs of motion, in which it has been excited by chemical changes. The ultimate cause of the heat produced is therefore to be found in these chemical changes.

By dissolving a metal in an acid, we produce an electrical current; this current, if passed through a wire, converts the wire into a magnet, by means of which many different effects may be produced. The cause of these phenomena is magnetism; the cause of the magnetic phenomena is to be found in the electrical current; and the ultimate cause of the electrical current is found to be a chemical change, a chemical action.

There are various causes by which force or motion may be produced. A bent spring, a current of air, the fall of water, fire applied to a boiler, the solution of a metal in an acid,—all these different

causes of motion may be made to produce the
same effect. But in the animal body we recognize
as the ultimate cause of all force only one cause,
the chemical action which the elements of the food
and the oxygen of the air mutually exercise on each
other. The only known ultimate cause of vital
force, either in animals or in plants, is a chemical
process. If this be prevented, the phenomena of
life do not manifest themselves, or they cease to be
recognizable by our senses. If the chemical action
be impeded, the vital phenomena must take new
forms.

According to the experiments of Despretz, 1 oz.
of carbon evolves, during its combustion, as much
heat as would raise the temperature of 105 oz. of
water at 32° to 167°, that is, by 135 degrees; in all,
therefore, 105 times 135°=14207 degrees of heat.
Consequently, the 13·9 oz. of carbon which are daily
converted into carbonic acid in the body of an
adult, evolve 13·9×14207°=197477·3 degrees of
heat. This amount of heat is sufficient to raise the
temperature of 1 oz. of water by that number of
degrees, or from 32° to 197509·3°; or to cause
136·8 lbs. of water at 32° to boil; or to heat 370 lbs.
of water to 98·3° (the temperature of the human
body); or to convert into vapour 24 lbs. of water
at 98·3°.

If we now assume that the quantity of water
vaporized through the skin and lungs in 24 hours
amounts to 48 oz. (3 lbs.), then there will remain,

after deducting the necessary amount of heat, 146380·4 degrees of heat, which are dissipated by radiation, by heating the expired air, and in the excrementitious matters.

In this calculation, no account has been taken of the heat evolved by the hydrogen of the food, during its conversion into water by oxidation within the body. But if we consider that the specific heat of the bones, of fat, and of the organs generally, is far less than that of water, and that consequently they require, in order to be heated to 98·3°, much less heat than an equal weight of water, no doubt can be entertained, that when all the concomitant circumstances are included in the calculation, the heat evolved in the process of combustion, to which the food is subjected in the body, is amply sufficient to explain the constant temperature of the body, as well as the evaporation from the skin and lungs.

VI. All experiments hitherto made on the quantity of oxygen which an animal consumes in a given time, and also the conclusions deduced from them as to the origin of animal heat, are destitute of practical value in regard to this question, since we have seen that the quantity of oxygen consumed varies according to the temperature and density of the air, according to the degree of motion, labour, or exercise, to the amount and quality of the food, to the comparative warmth of the clothing, and also according to the time within which the food is taken. Prisoners in the Bridewell at Marienschloss (a prison

where labour is enforced), do not consume more than
10·5 oz. of carbon daily; those in the House of
Arrest at Giessen, who are deprived of all exercise,
consume only 8·5 oz. ; (6) and in a family well known
to me, consisting of nine individuals, five adults, and
four children of different ages, the average daily
consumption of carbon for each, is not more than
9·5 oz. of carbon.* We may safely assume, as an ap-
proximation, that the quantities of oxygen consumed
in these different cases are in the ratio of these
numbers; but where the food contains meat, fat, and
wine, the proportions are altered by reason of the
hydrogen in these kinds of food which is oxidised,
and which, in being converted into water, evolves
much more heat for equal weights.

The attempts to ascertain the amount of heat
evolved in an animal for a given consumption of
oxygen have been equally unsatisfactory. Animals
have been allowed to respire in close chambers sur-
rounded with cold water; the increase of temperature in the water has been measured by the thermometer, and the quantity of oxygen consumed has
been calculated from the analysis of the air before

---

* In this family, the monthly consumption was 151 lbs. of
brown bread, 70 lbs. white bread, 132 lbs. meat, 19 lbs. sugar,
15·9 lbs. butter, 57 maass (about 24 gallons) of milk; the carbon of
the potatoes and other vegetables, of the poultry, game, and wine
consumed, having been reckoned as equal to that contained in
the excrementitious matters, the carbon of the above articles was
considered as being converted into carbonic acid.

and after the experiment. In experiments thus
conducted, it has been found that the animal lost
about $\frac{1}{10}$ more heat than corresponded to the oxygen
consumed ; and had the windpipe of the animal been
tied, the strange result would have been obtained of
a rise in the temperature of the water without any
consumption of oxygen. The animal was at the
temperature of 98° or 99°, and the water, in the
experiments of Despretz, was at 47·5°. Such experiments consequently prove, that when a great
difference exists between the temperature of the
animal body and that of the surrounding medium,
and when no motion is allowed, more heat is given
off than corresponds to the oxygen consumed. In
equal times, with free and unimpeded motion, a
much larger quantity of oxygen would be consumed
without a perceptible increase in the amount of
heat lost. The cause of these phenomena is obvious.
They appear naturally both in man and animals at
certain seasons of the year, and we say in such cases
that we are freezing, or experience the sensation of
cold.. It is plain, that if we were to clothe a man
in a metallic dress, and tie up his hands and feet,
the loss of heat, for the same consumption of oxygen,
would be far greater than if we were to wrap him
up in fur and woollen cloth. Nay, in the latter
case, we should see him begin to perspire, and warm
water would exude, in drops, through the finest
pores of his skin.

If to these considerations we add, that decisive

experiments are on record, in which animals were made to respire in an unnatural position, as for example, lying on the back, with the limbs tied so as to preclude motion, and that the temperature of their bodies was found to sink in a degree appreciable by the thermometer, we can hardly be at a loss what value we ought to attach to the conclusions drawn from such experiments as those above described.

These experiments and the conclusions deduced from them, in short, are incapable of furnishing the smallest support to the opinion that there exists, in the animal body, any other unknown source of heat, besides the mutual chemical action between the elements of the food and the oxygen of the air. The existence of the latter cannot be doubted or denied, and it is amply sufficient to explain all the phenomena.

VII. If we designate the production of force, the phenomena of motion in the animal body as *nervous life*, and the resistance, the condition of static equilibrium, as *vegetative life;* it is obvious that in all classes of animals the latter, namely, vegetative life, prevails over the former, nervous life, in the earlier stages of existence.

The passage or change of matter from a state of motion to a state of rest appears in an increase of the mass, and in the supply of waste; while the motion itself, or the production of force, appears in the shape of waste of matter.

In a young animal, the waste is less than the increase; and the female retains, up to a certain age, this peculiar condition of a more intense vegetative life. This condition does not cease in the female as in the male, with the complete developement of all the organs of the body.

The female in the lower animals, is, at certain seasons, capable of reproduction of the species. The vegetative life in her organism is rendered more intense by certain external conditions, such as temperature, food, &c.; the organism produces more than is wasted, and the result is the capacity of reproduction.

In the human species, the female organism is independent of those external causes which increase the intensity of vegetative life. When the organism is fully developed, it is at all times capable of reproduction of the species; and infinite wisdom has given to the female body the power, up to a certain age, of producing all parts of its organisation in greater quantity than is required to supply the daily waste.

This excess of production can be shewn to contain all the elements of a new organism, it is constantly accumulating, and is periodically expelled from the body, until it is expended in reproduction. This periodical discharge ceases when the ovum has been impregnated, and from this time every drop of the superabundant blood goes to produce an organism like that of the mother.

Exercise and labour cause a diminution in the

quantity of the menstrual discharge; and when it is suppressed in consequence of disease, the vegetative life is manifested in a morbid production of fat. When the equilibrium between the vegetative and nervous life is disturbed in the male, when, as in eunuchs, the intensity of the latter is diminished, the predominance of the former is shewn in the same form, in an increased deposit of fat.

VIII. If we hold, that increase of mass in the animal body, the developement of its organs, and the supply of waste,—that all this is dependant on the blood, that is, on the ingredients of the blood, then only those substances can properly be called nutritious or considered as food which are capable of conversion into blood. To determine, therefore, what substances are capable of affording nourishment, it is only necessary to ascertain the composition of the food, and to compare it with that of the ingredients of the blood.

Two substances require especial consideration as the chief ingredients of the blood; one of these separates immediately from the blood when withdrawn from the circulation. It is well known that in this case blood coagulates, and separates into a yellowish liquid, the *serum* of the blood, and a gelatinous mass, which adheres to a rod or stick in soft, elastic fibres, when coagulating blood is briskly stirred. This is the *fibrine* of the blood, which is identical in all its properties with muscular fibre, when the latter is purified from all foreign matters.

The second principal ingredient of the blood is contained in the serum, and gives to this liquid all the properties of the white of eggs, with which it is identical. When heated, it coagulates into a white elastic mass, and the coagulating substance is called *albumen*.

Fibrine and albumen, the chief ingredients of blood, contain, in all, seven chemical elements, among which nitrogen, phosphorus, and sulphur are found. They contain also the earth of bones. The serum retains in solution sea salt and other salts of potash and soda, in which the acids are carbonic, phosphoric, and sulphuric acids. The globules of the blood contain fibrine and albumen, along with a red colouring matter, in which iron is a constant element. Besides these, the blood contains certain fatty bodies in small quantity, which differ from ordinary fats in several of their properties.

Chemical analysis has led to the remarkable result, that fibrine and albumen contain the same organic elements united in the same proportion, so that two analyses, the one of fibrine and the other of albumen, do not differ more than two analyses of fibrine or two of albumen respectively do, in the composition of 100 parts.

In these two ingredients of blood the particles are arranged in a different order, as is shewn by the difference of their external properties; but in chemical composition, in the ultimate proportion of the organic elements, they are identical.

This conclusion has lately been beautifully confirmed by a distinguished physiologist (Dénis), who has succeeded in converting fibrine into albumen, that is, in giving it the solubility, and coagulability by heat, which characterize the white of egg.

Fibrine and albumen, besides having the same composition, agree also in this, that both dissolve in concentrated muriatic acid, yielding a solution of an intense purple colour. This solution, whether made with fibrine or albumen, has the very same re-actions with all substances yet tried.

Both albumen and fibrine, in the process of nutrition, are capable of being converted into muscular fibre, and muscular fibre is capable of being reconverted into blood. These facts have long been established by physiologists, and chemistry has merely proved that these metamorphoses can be accomplished under the influence of a certain force, without the aid of a third substance, or of its elements, and without the addition of any foreign element, or the separation of any element previously present in these substances.

If we now compare the composition of all organised parts with that of fibrine and albumen, the following relations present themselves :—

All parts of the animal body which have a decided shape, which form parts of organs, contain nitrogen. No part of an organ which possesses motion and life is destitute of nitrogen ; all of them contain likewise carbon and the elements of water, the latter, however, in no case in the proportion to form water.

The chief ingredients of the blood contain nearly 17 per cent. of nitrogen, and no part of an organ contains less than 17 per cent. of nitrogen. (7)

The most convincing experiments and observations have proved that the animal body is absolutely incapable of producing an elementary body, such as carbon or nitrogen, out of substances which do not contain it; and it obviously follows, that all kinds of food fit for the production either of blood, or of cellular tissue, membranes, skin, hair, muscular fibre, &c., must contain a certain amount of nitrogen, because that element is essential to the composition of the above-named organs; because the organs cannot create it from the other elements presented to them ; and, finally, because no nitrogen is absorbed from the atmosphere in the vital process.

The substance of the brain and nerves contains a large quantity of albumen, and, in addition to this, two peculiar fatty acids, distinguished from other fats by containing phosphorus (phosphoric acid ?). One of these contains nitrogen (Frémy).

Finally, water and common fat are those ingredients of the body which are destitute of nitrogen. Both are amorphous or unorganised, and only so far take part in the vital process as that their presence is required for the due performance of the vital functions. The inorganic constituents of the body are, iron, lime, magnesia, common salt, and the alkalies.

IX. The nutritive process in the carnivora is seen in its simplest form. This class of animals lives on the blood and flesh of the graminivora; but this blood and flesh is, in all its properties, identical with their own. Neither chemical nor physiological differences can be discovered.

The nutriment of carnivorous animals is derived originally from blood; in their stomach it becomes dissolved, and capable of reaching all other parts of the body; in its passage it is again converted into blood, and from this blood are reproduced all those parts of their organisation which have undergone change or metamorphosis.

With the exception of hoofs, hair, feathers, and the earth of bones, every part of the food of carnivorous animals is capable of assimilation.

In a chemical sense, therefore, it may be said that a carnivorous animal, in supporting the vital process, consumes itself. That which serves for its nutrition is identical with those parts of its organisation which are to be renewed.

The process of nutrition in graminivorous animals appears at first sight altogether different. Their digestive organs are less simple, and their food consists of vegetables, the great mass of which contains but little nitrogen.

From what substances, it may be asked, is the blood formed, by means of which their organs are developed? This question may be answered with certainty.

Chemical researches have shewn, that all such parts of vegetables as can afford nutriment to animals contain certain constituents which are rich in nitrogen; and the most ordinary experience proves that animals require for their support and nutrition less of these parts of plants in proportion as they abound in the nitrogenised constituents. Animals cannot be fed on matters destitute of these nitrogenised constituents.

These important products of vegetation are especially abundant in the seeds of the different kinds of grain, and of pease, beans, and lentils; in the roots and the juices of what are commonly called vegetables. They exist, however, in all plants, without exception, and in every part of plants in larger or smaller quantity.

These nitrogenised forms of nutriment in the vegetable kingdom may be reduced to three substances, which are easily distinguished by their external characters. Two of them are soluble in water, the third is insoluble.

When the newly-expressed juices of vegetables are allowed to stand, a separation takes place in a few minutes. A gelatinous precipitate, commonly of a green tinge, is deposited, and this, when acted on by liquids which remove the colouring matter, leaves a greyish white substance, well known to druggists as the deposit from vegetable juices. This is one of the nitrogenised compounds which serves for the nutrition of animals, and has been named *vegetable fibrine*. The juice of grapes is especially rich in this constituent, but it is most abundant in the seeds of wheat, and of the cerealia generally. It may be obtained from wheat flour by a mechanical operation, and in a state of tolerable purity; it is then called *gluten*, but the glutinous property belongs, not to vegetable fibrine, but to a foreign substance, present in small quantity, which is not found in the other cerealia.

The method by which it is obtained sufficiently proves that it is insoluble in water; although we cannot doubt that it was originally dissolved in the vegetable juice, from which it afterwards separated, exactly as fibrine does from blood.

The second nitrogenised compound remains dissolved in the juice after the separation of the fibrine. It does not separate from the juice at the ordinary temperature, but is instantly coagulated when the liquid containing it is heated to the boiling point.

When the clarified juice of nutritious vegetables, such as cauliflower, asparagus, mangel wurzel, or turnips, is made to boil, a coagulum is formed, which it is absolutely impossible to distinguish from the substance which separates as a coagulum, when the serum of blood or the white of an egg, diluted with water, are heated to the boiling point.

This is *vegetable albumen*. It is found in the greatest abundance in certain seeds, in nuts, almonds, and others, in which the starch of the gramineæ is replaced by oil.

The third nitrogenised constituent of the vegetable food of animals is *vegetable caseine*. It is chiefly found in the seeds of pease, beans, lentils, and similar leguminous seeds. Like vegetable albumen, it is soluble in water, but differs from it in this, that its solution is not coagulated by heat. When the solution is heated or evaporated, a skin forms on its surface, and the addition of an acid causes a coagulum, just as in animal milk.

These three nitrogenised compounds, vegetable fibrine, albumen, and caseine, are the true nitrogenised constituents of the food of graminivorous animals; all other nitrogenised compounds, occurring in plants, are either rejected by animals, as in the case of the characteristic principles of poisonous and medicinal plants, or else they occur in the food in such very small proportion, that they cannot possibly contribute to the increase of mass in the animal body.

The chemical analysis of these three substances has led to the very interesting result that they contain the same organic elements, united in the same proportion by weight; and, what is still more remarkable, that they are identical in composition with the chief constituents of blood, animal fibrine, and albumen. They all three dissolve in concen-

trated muriatic acid with the same deep purple
colour, and even in their physical characters, animal
fibrine and albumen are in no respect different from
vegetable fibrine and albumen. It is especially to
be noticed, that by the phrase, identity of composition, we do not here imply mere similarity, but that
even in regard to the presence and relative amount
of sulphur, phosphorus, and phosphate of lime, no
difference can be observed. (8)

How beautifully and admirably simple, with the
aid of these discoveries, appears the process of nutrition in animals, the formation of their organs,
in which vitality chiefly resides! Those vegetable
principles, which in animals are used to form blood,
contain the chief constituents of blood, fibrine and
albumen, ready formed, as far as regards their
composition. All plants, besides, contain a certain
quantity of iron, which re-appears in the colouring
matter of the blood. Vegetable fibrine and animal
fibrine, vegetable albumen and animal albumen,
hardly differ, even in form; if these principles be
wanting in the food, the nutrition of the animal is
arrested; and when they are present, the graminivorous animal obtains in its food the very same
principles on the presence of which the nutrition of
the carnivora entirely depends.

Vegetables produce in their organism the blood
of all animals, for the carnivora, in consuming the
blood and flesh of the graminivora, consume, strictly
speaking, only the vegetable principles which have

served for the nutrition of the latter. Vegetable fibrine and albumen take the same form in the stomach of the graminivorous animal as animal fibrine and albumen do in that of the carnivorous animal.

From what has been said, it follows that the developement of the animal organism and its growth are dependant on the reception of certain principles identical with the chief constituents of blood.

In this sense we may say that the animal organism gives to blood only its form ; that it is incapable of creating blood out of other substances which do not already contain the chief constituents of that fluid. We cannot, indeed, maintain that the animal organism has no power to form other compounds, for we know that it is capable of producing an extensive series of compounds, differing in composition from the chief constituents of blood ; but these last, which form the starting point of the series, it cannot produce.

The animal organism is a higher kind of vegetable, the developement of which begins with those substances, with the production of which the life of an ordinary vegetable ends. As soon as the latter has borne seed, it dies, or a period of its life comes to a termination.

In that endless series of compounds, which begins with carbonic acid, ammonia, and water, the sources of the nutrition of vegetables, and includes the most complex constituents of the animal brain,

E

there is no blank, no interruption. The first substance capable of affording nutriment to animals is the last product of the creative energy of vegetables.

The substance of cellular tissue and of membranes, of the brain and nerves, these the vegetable cannot produce.

The seemingly miraculous in the productive agency of vegetables disappears in a great degree, when we reflect that the production of the constituents of blood cannot appear more surprising than the occurrence of the fat of beef and mutton in cocoa beans, of human fat in olive oil, of the principal ingredient of butter in palm oil, and of horse fat and train oil in certain oily seeds.

X. While the preceding considerations leave little or no doubt as to the way in which the increase of mass in an animal, that is, its growth, is carried on, there is yet to be resolved a most important question, namely, that of the function performed in the animal system by substances containing no nitrogen, such as sugar, starch, gum, pectine, &c.

The most extensive class of animals, the graminivora, cannot live without these substances; their food must contain a certain amount of one or more of them, and if these compounds are not supplied, death quickly ensues.

This important inquiry extends also to the constituents of the food of carnivorous animals in the earliest periods of life; for this food also contains substances, which are not necessary for their support in the adult state.

The nutrition of the young of carnivora is obviously accomplished by means similar to those by which the graminivora are nourished; their developement is dependant on the supply of a fluid, which the body of the mother secretes in the shape of milk.

Milk contains only one nitrogenised constituent, known under the name of *caseine;* besides this, its chief ingredients are butter (fat), and sugar of milk.

The blood of the young animal, its muscular fibre, cellular tissue, nervous matter, and bones, must have derived their origin from the nitrogenised constituent of milk, the caseine; for butter and sugar of milk contain no nitrogen.

Now, the analysis of caseine has led to the result, which, after the details given in the last section, can hardly excite surprise, that this substance also is identical in composition with the chief constituents of blood, fibrine and albumen. Nay, more, a comparison of its properties with those of vegetable caseine has shewn that these two substances are identical in all their properties; insomuch, that certain plants, such as peas, beans, and lentils, are capable of producing the same substance which is formed from the blood of the mother, and employed in yielding the blood of the young animal. (9)

The young animal, therefore, receives, in the form

of caseine, which is distinguished from fibrine and albumen by its great solubility, and by not coagulating when heated, the chief constituent of the mother's blood. To convert caseine into blood no foreign substance is required, and in the conversion of the mother's blood into caseine, no elements of the constituents of the blood have been separated. When chemically examined, caseine is found to contain a much larger proportion of the earth of bones than blood does, and that in a very soluble form, capable of reaching every part of the body. Thus, even in the earliest period of its life, the developement of the organs, in which vitality resides, is, in the carnivorous animal, dependant on the supply of a substance, identical in organic composition with the chief constituents of its blood.

What, then, is the use of the butter and the sugar of milk ? How does it happen that these substances are indispensable to life ?

Butter and sugar of milk contain no fixed bases, no soda or potash. Sugar of milk has a composition closely allied to that of the other kinds of sugar, of starch, and of gum ; all of them contain carbon and the elements of water, the latter precisely in the proportion to form water.

There is added, therefore, by means of these compounds, to the nitrogenised constituents of food, a certain amount of carbon, or, as in the case of butter, of carbon and hydrogen ; that is, an excess of elements, which cannot possibly be employed in the

production of blood, because the nitrogenised substances contained in the food already contain exactly
the amount of carbon which is required for the production of fibrine and albumen.

The following considerations will shew that hardly
a doubt can be entertained, that this excess of carbon alone, or of carbon and hydrogen, is expended
in the production of animal heat, and serves to protect the organism from the action of the atmospheric
oxygen.

XI. In order to obtain a clearer insight into the
nature of the nutritive process in both the great
classes of animals, let us first consider the changes
which the food of the carnivora undergoes in their
organism.

If we give to an adult serpent, or boa constrictor,
a goat, a rabbit, or a bird, we find that the hair,
hoofs, horns, feathers, or bones of these animals, are
expelled from the body apparently unchanged. They
have retained their natural form and aspect, but
have become brittle, because of all their component
parts they have lost only that one which was capable
of solution, namely, the gelatine.   Fæces, properly
so called, do not occur in serpents any more than in
carnivorous birds.

We find, moreover, that, when the serpent has
regained its original weight, every other part of its
prey, the flesh, the blood, the brain, and nerves, in
short, every thing, has disappeared.

The only excrement, strictly speaking, is a sub-stance expelled by the urinary passage. When dry, it is pure white, like chalk; it contains much nitrogen, and a small quantity of carbonate and phosphate of lime mixed with the mass.

This excrement is urate of ammonia, a chemical compound, in which the nitrogen bears to the carbon the same proportion as in bicarbonate of ammonia. For every equivalent of nitrogen it contains two equivalents of carbon.

But muscular fibre, blood, membranes, and skin, contain four times as much carbon for the same amount of nitrogen, or eight equivalents to one; and if we add to this the carbon of the fat and nervous substance, it is obvious that the serpent has consumed, for every equivalent of nitrogen, much more than eight equivalents of carbon.

If now we assume that the urate of ammonia contains all the nitrogen of the animal consumed, then at least six equivalents of carbon, which were in combination with this nitrogen, must have been given out in a different form from the two equivalents which are found in the urate of ammonia.

Now we know, with perfect certainty, that this carbon has been given out through the skin and lungs, which could only take place in the form of an oxidised product.

The excrements of a buzzard which had been fed with beef, when taken out of the rectum, consisted, according to L. Gmelin and Tiedemann, of urate of

ammonia. In like manner, the fæces in lions and
tigers are scanty and dry, consisting chiefly of bone
earth, with mere traces of compounds containing
carbon ; but their urine contains, not urate of ammonia, but urea, a compound in which carbon and
nitrogen are to each other in the same ratio as in
neutral carbonate of ammonia.

Assuming that their food (flesh, &c.) contains
carbon and nitrogen in the ratio of eight equivalents
to one, we find these elements in their urine in the
ratio of one equivalent to one ; a smaller proportion
of carbon, therefore, than in serpents, in which respiration is so much less active.

The whole of the carbon and hydrogen which the
food of these animals contained, beyond the amount
which we find in their excrements, has disappeared,
in the process of respiration, as carbonic acid and
water.

Had the animal food been burned in a furnace,
the change produced in it would only have differed
in the form of combination assumed by the nitrogen
from that which it underwent in the body of the
animal. The nitrogen would have appeared, with
part of the carbon and hydrogen, as carbonate of
ammonia, while the rest of the carbon and hydrogen
would have formed carbonic acid and water. The
incombustible parts would have taken the form of
ashes, and any part of the carbon unconsumed from
a deficiency of oxygen would have appeared as
soot, or lamp-black. Now the solid excrements are

nothing else than the incombustible, or imperfectly
burned, parts of the food.

In the preceding pages it has been assumed that
the elements of the food are converted by the oxygen absorbed in the lungs into oxidised products;
the carbon into carbonic acid, the hydrogen into
water, and the nitrogen into a compound containing the same elements as carbonate of ammonia.

This is only true in appearance; the body, no
doubt, after a certain time, acquires its original
weight. The amount of carbon, and of the other
elements, is not found to be increased—exactly as
much carbon, hydrogen, and nitrogen has been given
out as was supplied in the food; but nothing is more
certain than that the carbon, hydrogen, and nitrogen given out, although equal in amount to what
is supplied in that form, do not directly proceed
from the food.

It would be utterly irrational to suppose that the
necessity of taking food, or the satisfying the appetite, had no other object than the production of
urea, uric acid, carbonic acid, and other excrementitious matters—of substances which the system
expels, and consequently applies to no useful purpose in the economy.

In the adult animal, the food serves to restore
the waste of matter; certain parts of its organs
have lost the state of vitality, have been expelled
from the substance of the organs, and have been

metamorphosed into new combinations, which are amorphous and unorganised.

The food of the carnivora is at once converted. into blood; out of the newly-formed blood those parts of organs which have undergone metamorphoses are reproduced. The carbon and nitrogen of the food thus become constituent parts of organs.

Exactly as much carbon and nitrogen is supplied to the organs by the blood, that is, ultimately, by the food, as they have lost by the transformations attending the exercise of their functions.

What then, it may be asked, becomes of the new compounds produced by the transformations of the organs, of the muscles, of the membranes and cellular tissue of the nerves and brain?

These new compounds cannot, owing to their solubility, remain in the situation where they are formed, for a well-known force, namely the circulation of the blood, opposes itself to this.

By the expansion of the heart, an organ in which two systems of tubes meet, which are ramified in a most minute network of vessels through all parts of the body, there is produced a vacuum, the immediate effect of which is, that all fluids which can penetrate into these vessels are urged with great force towards one side of the heart by the external pressure of the atmosphere. This motion is powerfully assisted by the contraction of the heart, alternating with its expansion, and caused by a force independent of the atmospheric pressure.

*

In a word, the heart is a forcing pump, which sends arterial blood into all parts of the body ; and also a suction pump, by means of which all fluids of whatever kind, as soon as they enter the absorbent vessels which communicate with the veins, are drawn towards the heart. This suction, arising from the vacuum caused by the expansion of the heart, is a purely mechanical act, which extends, as above stated, to fluids of every kind, to saline solutions, poisons, &c. It is obvious, therefore, that by the forcible entrance of arterial blood into the capillary vessels, the fluids contained in these, in other words, the soluble compounds formed by the transformations of organised parts, must be compelled to move towards the heart.

These compounds cannot be employed for the reproduction of those tissues from which they are derived. They pass through the absorbent and lymphatic vessels into the veins, where their accumulation would speedily put a stop to the nutritive process, were it not that this accumulation is prevented by two contrivances adapted expressly to this purpose, and which may be compared to filtering machines.

The venous blood, before reaching the heart, is made to pass through the liver ; the arterial blood, on the other hand, passes through the kidneys ; and these organs separate from both all substances incapable of contributing to nutrition.

Those new compounds which contain the nitrogen

of the transformed organs are collected in the urinary bladder, and being utterly incapable of any
further application in the system, are expelled from
the body.

Those, again, which contain the carbon of the
transformed tissues, are collected in the gall-bladder
in the form of a compound of soda, *the bile*, which is
miscible with water in every proportion, and which,
passing into the duodenum, mixes with the chyme.
All those parts of the bile which, during the digestive process, do not lose their solubility, return
during that process into the circulation in a state of
extreme division. The soda of the bile, and those
highly carbonised portions which are not precipitated
by a weak acid (together making $\frac{99}{100}$ths of the solid
contents of the bile), retain the capacity of resorption by the absorbents of the small and large intestines; nay, this capacity has been directly proved
by the administration of enemata containing bile,
the whole of the bile disappearing with the injected
fluid in the rectum.

Thus we know with certainty, that the nitrogenised compounds, produced by the metamorphosis of
organised tissues, after being separated from the
arterial blood by means of the kidneys are expelled
from the body as utterly incapable of further alteration ; while the compounds rich in carbon, derived
from the same source, return into the system of
carnivorous animals.

The food of the carnivora is identical with the

chief constituents of their bodies, and hence the
metamorphoses which their organs undergo must be
the same as those which, under the influence of the
vital force, take place in the matters which consti-
tute their food.

The flesh and blood consumed as food yield their
carbon for the support of the respiratory process,
while its nitrogen appears as uric acid, ammonia,
or urea.   But previously to these final changes, the
dead flesh and blood become living flesh and blood,
and it is, strictly speaking, the carbon of the compounds formed in the metamorphoses of living tissues that serves for the production of animal heat.

The food of the carnivora is converted into blood,
which is destined for the reproduction of organised
tissues; and by means of the circulation a current
of oxygen is conveyed to every part of the body.
The globules of the blood, which in themselves
can be shewn to take no share in the nutritive
process, serve to transport the oxygen, which they
give up in their passage through the capillary
vessels.   Here the current of oxygen meets with
the compounds produced by the transformation of
the tissues, and combines with their carbon to form
carbonic acid, with their hydrogen to form water.
Every portion of these substances which escapes this
process of oxidation is sent back into the circulation
in the form of the bile, which by degrees completely
disappears.

In the carnivora the bile contains the carbon of

the metamorphosed tissues; this carbon disappears in the animal body, and the bile likewise disappears in the vital process. Its carbon and hydrogen are given out through the skin and lungs as carbonic acid and water; and hence it is obvious that the elements of the bile serve for respiration and for the production of animal heat. Every part of the food of carnivorous animals is capable of forming blood; their excrements, excluding the urine, contain only inorganic substances, such as phosphate of lime; and the small quantity of organic matter which is found mixed with these is derived from excretions, the use of which is to promote their passage through the intestines, such as mucus. These excrements contain no bile and no soda; for water extracts from them no trace of any substance resembling bile, and yet bile is very soluble in water, and mixes with it in every proportion.

Physiologists can entertain no doubt as to the origin of the constituent parts of the urine and of the bile. When, from deprivation of food, the stomach contracts itself so as to resemble a portion of intestine, the gall-bladder, for want of the motion which the full stomach gives to it, cannot pour out the bile it contains; hence in animals starved to death we find the gall-bladder distended and full. The secretion of bile and of urine goes on during the winter sleep of hybernating animals; and we know that the urine of dogs, fed for three weeks exclusively on pure sugar, contains as much of the most highly nitrogenised constituent, urea, as in the normal condition. (Marchaud. Erdmaun's Journal für praktische Chemie, XIV. p. 495.)

Differences in the quantity of urea secreted in these and similar experiments are explained by the condition of the animal in regard to the amount of the natural motions permitted. Every motion increases the amount of organised tissue which undergoes metamorphosis. Thus, after a walk, the secretion of urine in man is invariably increased.

The urine of the mammalia, of birds, and of amphibia, contains uric acid or urea; and the excrements of the mollusca, and of insects, as of cantharides and of the butterfly of the silkworm, contain urate of ammonia. This constant occurrence of one or two nitrogenised compounds in the excretions of animals, while so great a difference exists in their food, clearly proves that these compounds proceed from one and the same source.

As little doubt can be entertained in regard to the function of the bile in the vital process. When we consider, that the acetate of potash, given in enema, or simply as a bath for the feet, renders the urine strongly alkaline (Rehberger in Tiedemann's Zeitschrift für Physiologie, II. 149), and that the change which the acetic acid here undergoes cannot be conceived without the addition of oxygen, it is obvious, that the soluble constituents of the bile, prone to change in a high degree as we know them to be, and which, as already stated, cannot be em-

ployed in the production of blood, must, when returned through the intestines into the circulation, in
like manner yield to the influence of the oxygen
which they meet. The bile is a compound of soda,
the elements of which, with the exception of the
soda, disappear in the body of a carnivorous animal.

In the opinion of many of the most distinguished
physiologists, the bile is intended solely to be excreted; and nothing is more certain, than that a
substance containing so very small a proportion of
nitrogen can have no share in the process of nutrition or reproduction of organised tissue. But
quantitative physiology must at once and decidedly
reject the opinion, that the bile serves no purpose
in the economy, and is incapable of further change.

No part of any organised structure contains soda;
only in the serum of the blood, in the fat of the
brain, and in the bile, do we meet with that alkali.
When the compounds of soda in the blood are converted into muscular fibre, membrane, or cellular
tissue, the soda they contain must enter into new
combinations. The blood which is transformed into
organised tissue gives up its soda to the compounds
formed by the metamorphoses of the previously
existing tissues. In the bile we find one of these
compounds of soda.

Were the bile intended merely for excretion, we
should find it, more or less altered, and also the
soda it contains, in the solid excrements. But,
with the exception of common salt, and of sulphate

of soda, which occur in all the animal fluids, only
mere traces of soda are to be found in the fæces.
The soda of the bile, therefore, at all events, must
have returned from the intestinal canal into the
organism, and the same must be true of the organic
matters which were in combination with it.

According to the observations of physiologists, a
man secretes daily from 17 to 24 oz. of bile ; a
large dog, 36 oz. ; a horse, 37 lbs. (Burdach's Physiologie, V. p. 260.) But the fæces of a man do
not on an average weigh more than $5\frac{1}{2}$ oz.; and
those of a horse $28\frac{1}{2}$ lbs., of which 21 lbs. are water,
and $7\frac{1}{2}$ lbs. dry fæces. (Boussingault.) The latter
yield to alcohol only $\frac{1}{20}$th part of their weight of
soluble matter.

If we assume the bile to contain 90 per cent. of
water, a horse secretes daily 592 oz. of bile, containing 59·2 oz. of solid matter; while $7\frac{1}{2}$ lbs. or
120 oz. of dried excrement yield only 6 oz. of matter soluble in alcohol, which might possibly be
bile.   But this matter is not bile ; when the alcohol is dissipated by evaporation, there remains a
soft, unctuous mass, altogether insoluble in water,
and which, when incinerated, leaves no alkaline
ashes, no soda. (10)

During the digestive process, therefore, the soda
of the bile, and, along with it, all the soluble parts
of that fluid, are returned into the circulation.   This
soda re-appears in the newly-formed blood, and,
finally, we find it in the urine in the form of phos-

phate, carbonate, and hippurate of soda. Berzelius found in 1,000 parts of fresh human fæces only nine parts of a substance similar to bile ; 5 ounces, therefore, would contain only 21 grains of dried bile, equivalent to 210 grains of fresh bile. But a man secretes daily from 9,640 to 11,520 grains of fluid bile, that is, from 45 to 56 times as much as can be detected in the matters discharged by the intestinal canal.

Whatever opinion we may entertain of the accuracy of the physiological experiments, in regard to the quantity of bile secreted by the different classes of animals ; thus much is certain, that even the maximum of the supposed secretion, in man and in the horse, does not contain as much carbon as is given out in respiration. With all the fat which is mixed with it, or enters into its composition, dried bile does not contain more than 69 per cent. of carbon. Consequently, if a horse secretes 37 lbs. of bile, this quantity will contain only 40 ounces of carbon. But the horse expires daily nearly twice as much in the form of carbonic acid. A precisely similar proportion holds good in man.

Along with the matter destined for the formation or reproduction of organs, the circulation conveys oxygen to all parts of the body. Now, into whatever combination the oxygen may enter in the blood, it must be held as certain, that such of the constituents of blood as are employed for reproduction, are not materially altered by it. In muscular fibre we find fibrine, with all the properties it had

F

in venous blood; the albumen in the blood does not combine with oxygen. The oxygen may possibly serve to convert into the gaseous state some unknown constituent of the blood; but those well-known constituents, which are employed in reproduction, cannot be destined to support the respiratory process; none of their properties can justify such an opinion.

Without attempting in this place to exhaust the whole question of the share taken by the bile in the vital operations, it follows, as has been observed, from the simple comparison of those parts of the food of the carnivora which are capable of assimilation, with the ultimate products into which it is converted, that all the carbon of the food, except that portion which is found in the urine, is given out as carbonic acid.

But this carbon was ultimately derived from the substance of the metamorphosed tissues; and this being admitted, the question of the necessity of substances containing much carbon and no nitrogen in the food of the young of the carnivora, and in that of the graminivora, is resolved in a strikingly simple manner.

XII. It cannot be disputed, that in an adult carnivorous animal, which neither gains nor loses weight, perceptibly, from day to day, its nourishment, the waste of organised tissue, and its consumption of oxygen, stand to each other in a well-defined and fixed relation.

The carbon of the carbonic acid given off, with that of the urine ; the nitrogen of the urine, and the hydrogen given off as ammonia and water ; these elements, taken together, must be exactly equal in weight to the carbon, nitrogen, and hydrogen of the metamorphosed tissues, and since these last are exactly replaced by the food, to the carbon, nitrogen, and hydrogen of the food. Were this not the case, the weight of the animal could not possibly remain unchanged.

But, in the young of the carnivora, the weight does not remain unchanged ; on the contrary, it increases from day to day by an appreciable quantity.

This fact presupposes, that the assimilative process in the young animal is more energetic, more intense, than the process of transformation in the existing tissues. If both processes were equally active, the weight of the body could not increase ; and were the waste by transformation greater, the weight of the body would decrease.

Now, the circulation in the young animal is not weaker, but, on the contrary, more rapid ; the respirations are more frequent ; and, for equal bulks, the consumption of oxygen must be greater rather than smaller in the young than in the adult animal. But, since the metamorphosis of organised parts goes on more slowly, there would ensue a deficiency of those substances, the carbon and hydrogen of which are adapted for combination with oxygen ; because, in the carnivora it is the new compounds,

produced by the metamorphosis of organised parts,
which nature has destined to furnish the necessary
resistance to the action of the oxygen, and to produce animal heat.   What is wanting for these purposes an infinite wisdom has supplied to the young
animal in its natural food.

The carbon and hydrogen of butter, and the carbon of the sugar of milk, no part of either of which
can yield blood, fibrine, or albumen, are destined
for the support of the respiratory process, at an age
when a greater resistance is opposed to the metamorphosis of existing organisms ; or, in other words,
to the production of compounds, which in the adult
state are produced in quantity amply sufficient for
the purpose of respiration.

The young animal receives the constituents of its
blood in the caseine of the milk.   A metamorphosis
of existing organs goes on, for bile and urine are
secreted ; the matter of the metamorphosed parts is
given off in the form of urine, of carbonic acid, and
of water ; but the butter and sugar of milk also
disappear ; they cannot be detected in the fæces.

The butter and sugar of milk are given out in the
form of carbonic acid and water, and their conversion into oxidised products furnishes the clearest
proof that far more oxygen is absorbed than is required to convert the carbon and hydrogen of the
metamorphosed tissues into carbonic acid and water.

The change and metamorphosis of organised tissues going on in the vital process in the young

animal, consequently yield, in a given time, much less carbon and hydrogen in the form adapted for the respiratory process than corresponds to the oxygen taken up in the lungs. The substance of its organised parts would undergo a more rapid consumption, and would necessarily yield to the action of the oxygen, were not the deficiency of carbon and hydrogen supplied from another source.

The continued increase of mass, or growth, and the free and unimpeded developement of the organs in the young animal, are dependent on the presence of foreign substances, which, in the nutritive process, have no other function than to protect the newly-formed organs from the action of the oxygen. It is the elements of these substances which unite with the oxygen; the organs themselves could not do so without being consumed; that is, growth, or increase of mass in the body, the consumption of oxygen remaining the same, would be utterly impossible.

The preceding considerations leave no doubt as to the purpose for which Nature has added to the food of the young of carnivorous mammalia substances devoid of nitrogen, which their organism cannot employ for nutrition, strictly so called, that is, for the production of blood; substances which may be entirely dispensed with in their nourishment in the adult state. In the young of carnivorous birds, the want of all motion is an obvious cause of diminished waste in the organised parts; hence, milk is not provided for them.

The nutritive process in the carnivora thus presents itself in two distinct forms; one of which we again meet with in the graminivora.

XIII. In the class of graminivorous animals, we observe, that during their whole life, their existence depends on the supply of substances having a composition identical with that of sugar of milk, or closely resembling it. Every thing that they consume as food contains a certain quantity of starch, or gum, or sugar, mixed with other matters.

The most abundant and widely-extended of the substances of this class is amylon or starch; it occurs in roots, seeds, and stalks, and even in wood, deposited in the form of roundish or oval globules, which differ from each other in size alone, being identical in chemical composition. (11) In the same plant, in the pea, for example, we find starch, the globules of which differ in size. Those in the expressed juice of the stalks have a diameter of from $\frac{1}{200}$ to $\frac{1}{150}$ of an inch, while those in the seed are three or four times larger. The globules in arrow-root and in potato starch are distinguished by their large size; those of rice and of wheat are remarkably small.

It is well known that starch may be converted into sugar by very different means. This change occurs in the process of germination, as in malting, and it is easily accomplished by the action of acids. The metamorphosis of starch into sugar depends simply, as is proved by analysis, on the addition of the elements of water. (12) All the carbon of the starch is found in the sugar ; none of its elements have been separated, and, except the elements of water, no foreign element has been added to it in this transformation.

In many, especially in pulpy fruits, which when unripe are sour and rough to the taste, but when ripe are sweet, as, for example, in apples and pears, the sugar is produced from the starch which the unripe fruit contains.

If we rub unripe apples or pears on a grater to a pulp, and wash this with cold water on a fine sieve, the turbid liquid which passes through deposits a very fine flour of starch, of which not even a trace can be detected in the ripe fruit. Many varieties become sweet while yet on the tree ; these are the summer or early apples and pears. Others, again, become sweet only after having been kept for a certain period after gathering. The after-ripening, as this change is called, is a purely chemical process, entirely independent of the vitality of the plant. When vegetation ceases, the fruit is capable of reproducing the species, that is, the kernel, stone, or true seed is fully ripe, but the fleshy covering from this period is subjected to the action of the atmosphere. Like all substances in a state of eremacausis, or decay, it absorbs oxygen, and gives off a certain quantity of carbonic acid gas.

In the same way as the starch in putrefying paste,

in which it is in contact with decaying gluten, is converted into sugar, the starch in the above-named
fruits, in a state of decay, or eremacausis, is transformed into grape sugar. The more starch the unripe fruit contains, the sweeter does it become when
ripe.

A close connection thus exists between sugar and
starch. By means of a variety of chemical actions,
which exert no other influence on the elements of
starch than that of changing the direction of their
mutual attraction, we can convert starch into sugar,.
but it is always grape sugar.

Sugar of milk in many respects resembles
starch ; (13) it is, by itself, incapable of the vinous
fermentation, but it acquires the property of resolv-
ing itself into alcohol and carbonic acid when it is
exposed to heat in contact with a substance in the
state of fermentation (such as putrefying cheese
in milk). In this case, it is first converted into
grape sugar; and it undergoes the same transformation, when it is kept in contact with acids—
with sulphuric acid, for example—at the ordinary
temperature.

Gum has the same composition in 100 parts as
cane sugar.(14) It is distinguished from the different
varieties of sugar by its not possessing the property
of being resolved into alcohol and carbonic acid by
the process of putrefaction. When placed in contact with fermenting substances, it undergoes no
appreciable change, whence we may conclude, with

some degree of probability, that its elements, in the peculiar arrangement according to which they are united, are held together with a stronger force than the elements of the different kinds of sugar.

There is, however, a certain relation between gum and sugar of milk, since both of them, when treated with nitric acid, yield the same oxidised product, namely, mucic acid, which cannot, under the same circumstances, be formed from any of the other kinds of sugar.

In order to shew more distinctly the similarity of composition in these different substances, which perform so important a part in the nutritive process of the graminivora, let us represent one equivalent of carbon by C (= 75·8), and one equivalent of water by *aqua* (= 112·4), we shall then have for the composition of these substances the following expressions :—

Starch ...... = 12 C + 10 aqua.
Cane Sugar... = 12 C + 10 aqua + 1 aqua.
Gum .... = 12 C + 10 aqua + 1 aqua.
Sugar of milk = 12 C + 10 aqua + 2 aqua.
Grape Sugar = 12 C + 10 aqua + 4 aqua.

For the same number of equivalents of carbon, starch contains 10 equivalents, cane-sugar and gum 11 equivalents, sugar of milk 12 equivalents, and grape-sugar 14 equivalents, of water, or the elements of water.

XIV. In these different substances, some one of which is never wanting in the food of the gramnivora, there is added to the nitrogenised constituents of this food, to the vegetable albumen, fibrine, and caseine, from which their blood is formed, strictly speaking, only a certain excess of carbon, which the animal organism cannot possibly employ to produce fibrine or albumen, because the nitrogenised constituents of the food already contain the carbon necessary for the production of blood, and because the blood in the body of the carnivora is formed without the aid of this excess of carbon.

The function performed in the vital process of the graminivora by these substances (sugar, gum, &c.) is indicated in a very clear and convincing manner, when we take into consideration the very small relative amount of the carbon which these animals consume in the nitrogenised constituents of their food, which bears no proportion whatever to the oxygen absorbed through the skin and lungs.

A horse, for example, can be kept in perfectly good condition, if he obtain as food 15 lbs. of hay and $4\frac{1}{2}$ lbs. of oats, daily. If we now calculate the whole amount of nitrogen in these matters, as ascertained by analysis ($1\cdot5$ per cent. in the hay, $2\cdot2$ per cent. in the oats), (15) in the form of blood, that is, as fibrine and albumen, with the due proportion of water in blood (80 per cent.), the horse receives daily no more than $4\frac{1}{2}$ oz. of nitrogen, corresponding to about 8 lbs. of blood. But along with this nitrogen, that is, combined with it in the form of fibrine or albumen, the animal receives only about $14\frac{1}{2}$ oz. of carbon. Only about 8 oz. of this can be employed to support respiration, for with the nitrogen expelled in the urine there are combined, in the form of urea, 3 oz., and in the form of hippuric acid, $3\frac{1}{2}$ oz., of carbon.

Without going further into the calculation, it will readily be admitted, that the volume of air inspired and expired by a horse, the quantity of oxygen consumed, and, as a necessary consequence, the amount of carbonic acid given out by the animal, is much greater than in the respiratory process in man. But an adult man consumes daily about 14 oz. of carbon, and the determination of Boussingault, according to which a horse expires 79 oz. daily, cannot be very far from the truth.

In the nitrogenised constituents of his food, therefore, the horse receives rather less than the fifth part of the carbon which his organism requires for the support of the respiratory process; and we see that the wisdom of the Creator has added to his food the $\frac{4}{5}$ths which are wanting, in various forms, as, starch, sugar, &c. with which the animal must be supplied, or his organism will be destroyed by the action of the oxygen.

It is obvious, that in the system of the graminivora, whose food contains so small a proportion, relatively, of the constituents of blood, the process of metamorphosis in existing tissues, and consequently their restoration or reproduction, must go on far less rapidly than in the carnivora. Were this not the case, a vegetation a thousand times more luxuriant than the actual one would not suffice for their nourishment. Sugar, gum, and starch, would no longer be necessary to support life in these animals, because, in that case, the products of the waste, or metamorphosis of the organised tissues, would contain enough of carbon to support the respiratory process.

Man, when confined to animal food, requires for his support and nourishment extensive sources of food, even more widely extended than the lion and tiger, because, when he has the opportunity, he kills without eating.

A nation of hunters, on a limited space, is utterly incapable of increasing its numbers beyond a certain point, which is soon attained. The carbon necessary for respiration must be obtained from the animals, of which only a limited number can live on the space supposed. These animals collect from plants the constituents of their organs and of their blood, and yield them, in turn, to the savages who live by the chase alone. They, again, receive this food unaccompanied by those compounds, destitute of nitrogen, which, during the life of the animals, served to support the respiratory process. In such men, confined to an animal diet, it is the carbon of the flesh and of the blood which must take the place of starch and sugar.

But 15 lbs. of flesh contain not more carbon than

4 lbs. of starch, (16) and while the savage with one animal and an equal weight of starch could maintain life
and health for a certain number of days, he would
be compelled, if confined to flesh, in order to procure the carbon necessary for respiration, during the
same time, to consume five such animals.

It is easy to see, from these considerations, how
close the connection is between agriculture and the
multiplication of the human species. The cultivation
of our crops has ultimately no other object than the
production of a maximum of those substances which
are adapted for assimilation and respiration, in the
smallest possible space. Grain and other nutritious
vegetables yield us, not only in starch, sugar, and
gum, the carbon which protects our organs from the
action of oxygen, and produces in the organism the
heat which is essential to life, but also in the form
of vegetable fibrine, albumen, and caseine, our
blood, from which the other parts of our body are
developed.

Man, when confined to animal food, respires, like
the carnivora, at the expense of the matters produced by the metamorphosis of organised tissues;
and, just as the lion, tiger, hyæna, in the cages of a
menagerie, are compelled to accelerate the waste of
the organised tissues by incessant motion, in order to
furnish the matter necessary for respiration, so, the
savage, for the very same object, is forced to make
the most laborious exertions, and go through a vast
amount of muscular exercise. He is compelled to

consume force merely in order to supply matter for respiration.

Cultivation is the economy of force. Science teaches us the simplest means of obtaining the greatest effect with the smallest expenditure of power, and with given means to produce a maximum of force. The unprofitable exertion of power, the waste of force in agriculture, in other branches of industry, in science, or in social economy, is characteristic of the savage state, or of the want of cultivation.

XV. A comparison of the urine of the carnivora with that of the graminivora shews very clearly, that the process of metamorphosis in the tissues is different, both in form and in rapidity, in the two classes of animals.

The urine of carnivorous animals is acid, and contains alkaline bases united with uric, phosphoric, and sulphuric acids. We know perfectly the source of the two latter acids. All the tissues, with the exception of cellular tissue and membrane, contain phosphoric acid and sulphur, which latter element is converted into sulphuric acid by the oxygen of the arterial blood. In the various fluids of the body there are only traces of phosphates or sulphates, except in the urine, where both are found in abundance. It is plain that they are derived from the metamorphosed tissues; they enter into the venous blood in the form of soluble salts, and are separated from it in its passage through the kidneys.

The urine of the graminivora is alkaline; it contains alkaline carbonates in abundance, and so small a portion of alkaline phosphates as to have been overlooked by most observers.

The deficiency or absence of alkaline phosphates in the urine of the graminivora, obviously indicates the slowness with which the tissues in this class of animals are metamorphosed; for if we assume that a horse consumes a quantity of vegetable fibrine and albumen corresponding to the amount of nitrogen in his daily food (about $4\frac{1}{2}$ oz.), and that the quantity of tissue metamorphosed is equal to that newly formed, then the quantity of phosphoric acid which on these suppositions would exist in the urine is not so small as not to be easily detected by analysis in the daily secretion of urine (3 lbs. according to Boussingault); for it would amount to 0·8 per cent. But, as above stated, most observers have been unable to detect phosphoric acid in the urine of the horse.

Hence it is obvious, that the phosphoric acid, which in consequence of the metamorphosis of tissues is produced in the form of soluble alkaline phosphates, must re-enter the circulation in this class of animals. It is there employed in forming brain and nervous matter, to which it is essential, and also, no doubt, in contributing to the supply of the earthy part of the bones. It is probable, however, that

*

the greater part of the earth of bones is obtained by the direct assimilation of phosphate of lime, while the soluble phosphates are better adapted for the production of nervous matter.

In the graminivora, therefore, whose food contains so small a proportion of phosphorus or of phosphates, the organism collects all the soluble phosphates produced by the metamorphosis of tissues, and employs them for the developement of the bones and of the phosphorised constituents of the brain; the organs of excretion do not separate these salts from the blood. The phosphoric acid which, by the change of matter, is separated in the uncombined state, is not expelled from the body as phosphate of soda; but we find it in the solid excrements in the form of insoluble earthy phosphates.

XVI. If we now compare the capacity for increase of mass, the assimilative power in the graminivora and carnivora, the commonest observations indicate a very marked difference.

A spider, which sucks with extreme voracity the blood of the first fly, is not disturbed or excited by a second or third. A cat will eat the first, and perhaps the second mouse presented to her, but even if she kills a third, she does not devour it. Exactly similar observations have been made in regard to lions and tigers, which only devour their prey when urged by hunger. Carnivorous animals, indeed, require less food for their mere support, because their skin is destitute of perspiratory pores, and because they consequently lose, for equal bulks, much less heat than graminivorous animals, which are compelled to restore the lost heat by means of food adapted for respiration.

How different is the energy and intensity of vegetative life in the graminivora. A cow, or a sheep, in the meadow, eats, almost without interruption, as long as the sun is above the horizon. Their system possesses the power of converting into organised tissues all the food they devour beyond the quantity required for merely supplying the waste of their bodies.

All the excess of blood produced is converted into cellular and muscular tissue; the graminivorous animal becomes fleshy and plump, while the flesh of the carnivorous animal is always tough and sinewy.

If we consider the case of a stag, a roe-deer, or a hare, animals which consume the same food as cattle and sheep, it is evident that, when well supplied with food, their growth in size, their fattening, must depend on the quantity of vegetable albumen, fibrine, or caseine, which they consume. With free and unimpeded motion and exercise, enough of oxygen is absorbed to consume the carbon of the gum, sugar, starch, and of all similar soluble constituents of their food.

But all this is very differently arranged in our domestic animals, when, with an abundant supply

of food, we check the processes of cooling and ex-
halation, as we do when we feed them in stables,
where free motion is impossible.

The stall-fed animal eats, and reposes merely for
digestion. It devours in the shape of nitrogenised
compounds far more food than is required for reproduction, or the supply of waste alone; and at the
same time it eats far more of substances devoid of
nitrogen than is necessary merely to support respiration and to keep up animal heat. Want of
exercise and diminished cooling are equivalent to a
deficient supply of oxygen; for when these circumstances occur, the animal absorbs much less oxygen
than is required to convert into carbonic acid the
carbon of the substances destined for respiration.
Only a small part of the excess of carbon thus occa-
sioned is expelled from the body in the horse and
ox, in the form of hippuric acid; and all the remainder is employed in the production of a substance
which, in the normal state, only occurs in small
quantity as a constituent of the nerves and brain.
This substance is *fat*.

In the normal condition, as to exercise and labour,
the urine of the horse and ox contains benzoic acid
(with 14 equivalents of carbon); but as soon as the
animal is kept quiet in the stable, the urine contains
hippuric acid (with 18 equivalents of carbon).

The flesh of wild animals is devoid of fat; while
that of stall-fed animals is covered with that substance. When the fattened animal is allowed to

move more freely in the air, or compelled to draw
heavy burdens, the fat again disappears.

It is evident, therefore, that the formation of fat
in the animal body is the result of a want of due
proportion between the food taken into the stomach
and the oxygen absorbed by the lungs and the skin.

A pig, when fed with highly nitrogenised food,
becomes full of flesh; when fed with potatoes
(starch) it acquires little flesh, but a thick layer of
fat. The milk of a cow, when stall-fed, is very rich
in butter, but in the meadow is found to contain
more caseine, and in the same proportion less butter
and sugar of milk. In the human female, beer and
farinaceous diet increase the proportion of butter
in the milk; an animal diet yields less milk, but it
is richer in caseine.

If we reflect, that in the entire class of carnivora,
the food of which contains no substance devoid of
nitrogen except fat, the production of fat in the body
is utterly insignificant; that even in these animals,
as in dogs and cats, it increases as soon as they live
on a mixed diet; and that we can increase the formation of fat in other domestic animals at pleasure, but
only by means of food containing no nitrogen; we
can hardly entertain a doubt that such food, in its
various forms of starch, sugar, &c., is closely connected with the production of fat.

In the natural course of scientific research, we
draw conclusions from the food in regard to the
tissues or substances formed from it; from the ni-

trogenised constituents of plants we draw certain inferences as to the nitrogenised constituents of the blood; and it is quite in accordance with this, the natural method, that we should seek to establish the relations of those parts of our food which are devoid of nitrogen and those parts of the body which contain none of that element. It is impossible to overlook the very intimate connection between them.

If we compare the composition of sugar of milk, of starch, and of the other varieties of sugar, with that of mutton and beef suet and of human fat, we find that in all of them the proportion of carbon to hydrogen is the same, and that they only differ in that of oxygen.

According to the analyses of Chevreul, mutton fat, human fat, and hog's lard contain 79 per cent. of carbon to 11·1, 11·4, and 11·7 per cent. of hydrogen respectively. (16)

Starch contains 44·91 carbon to 6·11 hydrogen
Gum and sugar 42·58 —— to 6·37 ditto. (17)

It is obvious that these numbers, representing the relative proportions of carbon and hydrogen in starch, gum, and sugar, are in the same ratio as the carbon and hydrogen in the different kinds of fat; for

$$44·91 : 6·11 = 79 : 10·99$$
$$42·58 : 6·37 = 79 : 11·80$$

From which it follows, that sugar, starch, and gum, by the mere separation of a part of their oxygen, may pass into fat, or at least into a substance having exactly the composition of fat. If from the formula of starch, $C_{12}H_{10}O_{10}$, we take 9 equivalents of oxygen, there will remain in 100 parts—

C_{12} .... 79·4
H_{10} .... 10·8
O .... 9·8

The empirical formula of fat which comes nearest to this is $C_{11}H_{10}O$, which gives in 100 parts—

C_{11} .... 78·9
H_{10} .... 11·6
O 9.5

According to this formula, an equivalent of starch, in order to be changed into fat, would lose 1 equivalent of carbonic acid, CO_2, and 7 equivalents of oxygen.

Now the composition of all saponifiable fatty bodies agrees very closely with one or other of these two formulæ.

If from 3 equivalents of sugar of milk, $3C_{12}H_{12}O_{12}$ $= C_{36}H_{36}O_{36}$, we take away four equivalents of water and 31 of oxygen, there will remain $C_{36}H_{22}O$, a formula which accurately represents the composition of cholesterine, the fat of bile. (18)

Whatever views we may entertain regarding the origin of the fatty constituents of the body, this much at least is undeniable, that the herbs and roots consumed by the cow contain no butter; that in hay or the other fodder of oxen no beef suet exists; that no hog's lard can be found in the potato refuse given to swine; and that the food of geese or fowls contains no goose fat or capon fat. The masses of

fat found in the bodies of these animals are formed
in their organism; and when the full value of this
fact is recognized, it entitles us to conclude that a
certain quantity of oxygen, in some form or other,
separates from the constituents of their food; for
without such a separation of oxygen, no fat could
possibly be formed from any one of these substances.

The chemical analysis of the constituents of the
food of the graminivora shews in the clearest manner that they contain carbon and oxygen in certain
proportions; which, when reduced to equivalents,
yield the following series :—

In vegetable fibrine, albumen, and caseine, there are con-
    tained, for...................... 120 eq. carbon, 36 eq. oxygen
In starch 120 100
In cane sugar 120 110
In gum 120 110
In sugar of milk 120 120
In grape sugar 120 140

*Now in all fatty bodies there are contained, on an
average—*

    for............................. 120 eq. carb. only 10 eq. oxygen.

Since the carbon of the fatty constituents of the
animal body is derived from the food, seeing that
there is no other source whence it can be derived,
it is obvious, if we suppose fat to be formed from
albumen, fibrine, or caseine, that, for every 120 equivalents of carbon deposited as fat, 26 equivalents of
oxygen must be separated from the elements of these
substances; and further, if we conceive fat to be

formed from starch, sugar, or sugar of milk, that for the same amount of carbon there must be separated 90, 100, and 110 equivalents of oxygen from these compounds respectively.

There is, therefore, but one way in which the formation of fat in the animal body is possible, and this is absolutely the same in which its formation in plants takes place ; it is a separation of oxygen from the elements of the food.

The carbon which we find deposited in the seeds and fruits of vegetables, in the form of oil and fat, was previously a constituent of the atmosphere, and was absorbed by the plant as carbonic acid. Its conversion into fat was accomplished under the influence of light, by the vital force of the vegetable; and the greater part of the oxygen of this carbonic acid was returned to the atmosphere as oxygen gas.*

In contradistinction to this phenomenon of vitality in plants, we know that the animal system absorbs oxygen from the atmosphere, and that this oxygen is again given out in combination with carbon or hydrogen ; we know, that in the formation of carbonic acid and water, the heat necessary to sustain the constant temperature of the body is produced, and that a process of oxidation is the only source of animal heat.

Whether fat be formed by the decomposition of

* See Appendix, No. 19, on the formation of wax and honey by the bee.

fibrine and albumen, the chief constituents of blood, or by that of starch, sugar, or gum, this decomposition must be accompanied by the separation of oxygen from the elements of these compounds. But this oxygen is not given out in the free state, because it meets in the organism with substances possessing the property of entering into combination with it. In fact, it is given out in the same forms as that which is absorbed from the atmosphere by the skin and lungs.

It is easy to see, from the above considerations, that a very remarkable connection exists between the formation of fat and the respiratory process.

XVIII. The abnormal condition, which causes the deposit of fat in the animal body, depends, as was formerly stated, on a disproportion between the quantity of carbon in the food and that of oxygen absorbed by the skin and lungs. In the normal condition, the quantity of carbon given out is exactly equal to that which is taken in the food, and the body acquires no increase of weight from the accumulation of substances containing much carbon and no nitrogen.

If we increase the supply of highly carbonised food, then the normal state can only be preserved on the condition that, by exercise and labour, the waste of the body is increased, and the supply of oxygen augmented in the same proportion.

The production of fat is always a consequence of

a deficient supply of oxygen, for oxygen is absolutely indispensable for the dissipation of the excess of carbon in the food. This excess of carbon, deposited in the form of fat, is never seen in the Bedouin or in the Arab of the desert, who exhibits with pride to the traveller his lean, muscular, sinewy limbs, altogether free from fat; but in prisons and jails it appears as a puffiness in the inmates, fed, as they are, on a poor and scanty diet; it appears in the sedentary females of oriental countries; and finally, it is produced under the well-known conditions of the fattening of domestic animals.

The formation of fat depends on a deficiency of oxygen; but in this process, in the formation of fat itself, there is opened up a new source of oxygen, a new cause of animal heat.

The oxygen set free in the formation of fat is given out in combination with carbon or hydrogen; and whether this carbon and hydrogen proceed from the substance that yields the oxygen, or from other compounds, still there must have been generated by this formation of carbonic acid or water as much heat as if an equal weight of carbon or hydrogen had been burned in air or in oxygen gas.

If we suppose that from 2 equivalents of starch 18 equivalents of oxygen are disengaged, and that these 18 equivalents of oxygen combine with 9 equivalents of carbon, from the bile, for example, no one can doubt that, in this case, exactly as much heat must be developed, as if these 9 equivalents of carbon had been directly burned. In this form, therefore, the disengagement of heat as a consequence of the formation of fat would be undeniable; and it could only be considered hypothetical, on the supposition that carbon and oxygen were disengaged from one and the same substance, in the proportions to yield carbonic acid.

If, for example, we suppose that from 2 atoms of starch, $C_{24}H_{20}O_{20}$, the elements of 9 equivalents of carbonic acid are separated, there will remain a compound containing, for 15 equivalents of carbon, 20 of hydrogen and 2 of oxygen; for

$$C_{24}H_{20}O_{20} = C_9O_{18} + C_{15}H_{20}O_2.$$

Or, if we assume that oxygen is separated from starch in the form both of carbonic acid and water, then, after subtracting the elements of 6 equivalents of water and 6 of carbonic acid, there would remain the compound $C_{18}H_{14}O_2$; for

$$C_{24}H_{20}O_{20} = C_6O_{12} + H_6O_6 + C_{18}H_{14}O_2.$$

Assuming, then, the separation of oxygen in either of these forms, it remains to be decided whether the carbonic acid and water given off were contained, *as such*, in the starch, or not.

If they were ready formed in the starch, the separation might occur without the disengagement of heat; but if the carbon and hydrogen were present in any other form in the starch (or in the compound from which the fat was produced), it is obvious that a change in the arrangement of the atoms must have occurred, in consequence of which the atoms of the carbon and of the hydrogen have united with those of the oxygen, to form carbonic acid and water.

Now, so far as chemical researches have gone, our knowledge of the constitution of starch, and of the varieties of sugar, will justify no other conclusion than this, that these substances contain *no ready formed carbonic acid.*

We are acquainted with a large number of processes of metamorphosis of a similar kind, in which the elements of carbonic acid and water are separated from certain pre-existing compounds ; and we know with certainty that all these processes are accompanied by a disengagement of heat, exactly as if the carbon and hydrogen combined directly with oxygen.

Such a disengagement of carbonic acid, for example, occurs in all processes of fermentation or putrefaction, which are, without exception, accompanied with the generation of heat.

In the fermentation of a saccharine solution, in consequence of a new arrangement of the elements of the sugar, a certain part of its carbon and oxygen unite to form carbonic acid, which separates as gas ; and as another result of this decomposition, we obtain a volatile combustible liquid, containing little oxygen, namely, alcohol.

If we add to 2 equivalents of sugar the elements of 12 equivalents of water, and subtract from the sum of the atoms 24 equivalents of oxygen, there remain 6 equivalents of alcohol.

$$(C_{24}H_{24}O_{24} + H_{12}O_{12}) - O_{24} = C_{24}H_{36}O_{12} = 6 \text{ eq. alcohol.}$$

These 24 equivalents of oxygen suffice to oxidise completely a third equivalent of sugar—that is, to convert its carbon into carbonic acid and its hydrogen into water, and by this oxidation we recover the 12 equivalents of water supposed to be added in the former part of the process, exactly as if this water had taken no share in it.

$$C_{12}H_{12}O_{12} + O_{24} = 12CO_2 + 12HO.$$

According to the ordinary view, 12 equivalents of carbonic acid separate from 3 of sugar, yielding 6 of alcohol—that is, exactly the same amount of these products as if two-thirds of the sugar had yielded oxygen to the remaining third, so as completely to oxidise its elements.

$$C_{36}H_{36}O_{36} = C_{24}H_{36}O_{12} + 12CO_2.*$$

By a comparison of these two methods of representing the same change, it will easily be seen that the division or splitting of a compound like sugar into carbonic acid, on the one hand, and a compound containing little oxygen, on the other, is in its results perfectly equivalent to a separation of oxygen from a certain portion of the compound and the oxidation or combustion of another portion of it at the expense of this oxygen.

It is well known that the temperature of a fermenting liquid rises ; and if we assume that a hogshead of wort, holding 1,200 litres = 2,400 lbs.,

---

* For an explanation of the formulæ and equations employed, see the Introduction to the Appendix.

French weight, contains 16 per cent. of sugar, in all
384 lbs., then, during the fermentation of this sugar,
an amount of heat must be generated equal to that
which would be produced by the combustion of
51 lbs. of carbon.

This is equal to a quantity of heat by which
every pound of the liquid might be heated by
297·9°; that is, supposing the decomposition of
the sugar to occur in a period of time too short
to be measured. This is well known not to be
the case; the fermentation lasts five or six days,
and each pound of liquid receives the 297·9 degrees of heat during a period of 120 hours. In
each hour there is, therefore, set free an amount
of heat capable of raising the temperature of each
pound of liquid 1·4 degree; a rise of temperature which is very powerfully counteracted by external . cooling and by the vaporization of alcohol
and water.

The formation of fat, like other analogous phenomena in which oxygen is separated in the form
of carbonic acid, is consequently accompanied by a
disengagement of heat. This change supplies to
the animal body a certain proportion of the oxygen
indispensable to the vital processes ; and this especially in those cases in which the oxygen absorbed
by the skin and lungs is not sufficient to convert
into carbonic acid the whole of the carbon adapted
for this combination.

*

This excess of carbon, as it cannot be employed to form a part of any organ, is deposited in the cellular tissue in the form of tallow or oil.

At every period of animal life, when there occurs a disproportion between the carbon of the food and the inspired oxygen, the latter being deficient, fat must be formed. Oxygen separates from existing compounds, and this oxygen is given out as carbonic acid or water. The heat generated in the formation of these two products contributes to keep up the temperature of the body.

Every pound of carbon which obtains the oxygen necessary to convert it into carbonic acid from substances which thereby pass into fat, must disengage as much heat as would raise the temperature of 200 lbs. of water by 70°,—that is, from 32° to 102°.

Thus, in the formation of fat, the vital force possesses a means of counteracting a deficiency in the supply of oxygen, and consequently in that of the heat indispensable for the vital process.

Experience teaches us that in poultry, the maximum of fat is obtained by tying the feet, and by a medium temperature. These animals in such circumstances may be compared to a plant possessing in the highest degree the power of converting all food into parts of its own structure. The excess of the constituents of blood forms flesh and other organised tissues, while that of starch, sugar, &c.,

is converted into fat. When animals are fattened on food destitute of nitrogen, only certain parts of their structure increase in size. Thus, in a goose, fattened in the method above alluded to, the liver becomes three or four times larger than in the same animal, when well fed with free motion, while we cannot say that the organised structure of the liver is thereby increased. The liver of a goose fed in the ordinary way is firm and elastic; that of the imprisoned animal is soft and spongy. The difference consists in a greater or less expansion of its cells, which are filled with fat.

In some diseases, the starch, sugar, &c., of the food obviously do not undergo the changes which enable them to assist in respiration, and consequently to be converted into fat. Thus, in diabetes mellitus, the starch is only converted into grape sugar, which is expelled from the body without further change.

In other diseases, as for example in inflammation of the liver, we find the blood loaded with fat and oil; and in the composition of the bile there is nothing at all inconsistent with the supposition that some of its constituents may be transformed into fat.

XIX. According to what has been laid down in the preceding pages, the substances of which the food of man is composed may be divided into two classes; into *nitrogenised* and *non-nitrogenised*. The former are capable of conversion into blood; the latter incapable of this transformation.

Out of those substances which are adapted to the formation of blood are formed all the organised tissues. The other class of substances, in the normal state of health, serve to support the process of respiration. The former may be called the *plastic elements of nutrition*; the latter, *elements of respiration.* Among the former we reckon—

Vegetable fibrine.
Vegetable albumen.
Vegetable caseine.
Animal flesh.
Animal blood.

Among the elements of respiration in our food, are—

Fat.   Pectine.
Starch.   Bassorine.
Gum.   Wine.
Cane Sugar.   Beer.
Grape Sugar. Spirits.
Sugar of milk.

XX. The most recent and exact researches have established as a universal fact, to which nothing yet known is opposed, that the nitrogenised constituents of vegetable food have a composition identical with that of the constituents of the blood.

No nitrogenised compound, the composition of which differs from that of fibrine, albumen, and caseine, is capable of supporting the vital process in animals.

The animal organism unquestionably possesses the

power of forming, from the constituents of its blood, the substance of its membranes and cellular tissue, of the nerves and brain, of the organic part of cartilages and bones. But the blood must be supplied to it ready formed in every thing but its form—that is, in its chemical composition. If this be not done, a period is rapidly put to the formation of blood, and consequently to life.

This consideration enables us easily to explain how it happens that the tissues yielding gelatine or chondrine, as, for example, the gelatine of skin or of bones, are not adapted for the support of the vital process; for their composition is different from that of fibrine or albumen. It is obvious that this means nothing more than that those parts of the animal organism which form the blood do not possess the power of effecting a transformation in the arrangement of the elements of gelatine, or of those tissues which contain it. The gelatinous tissues, the gelatine of the bones, the membranes, the cells, and the skin, suffer, in the animal body, under the influence of oxygen and moisture, a progressive alteration; a part of these tissues is separated, and must be restored from the blood; but this alteration and restoration is obviously confined within very narrow limits.

While, in the body of a starving or sick individual, the fat disappears, and the muscular tissue takes once more the form of blood, we find that the tendons and membranes retain their natural condition;

H   *

the limbs of the dead body retain their connections, which depend on the gelatinous tissues.

On the other hand, we see that the gelatine of bones devoured by a dog entirely disappears, while only the bone earth is found in his excrements. The same is true of man, when fed on food rich in gelatine, as, for example, strong soup. The gelatine is not to be found either in the urine or in the fæces, and consequently must have undergone a change, and must have served some purpose in the animal economy. It is clear, that the gelatine must be expelled from the body in a form different from that in which it was introduced as food.

When we consider the transformation of the albumen of the blood into a part of an organ composed of fibrine, the identity in composition of the two substances renders the change easily conceivable. Indeed we find the change of a dissolved substance into an insoluble organ of vitality, chemically speaking, natural and easily explained, on account of this very identity of composition. Hence the opinion is not unworthy of a closer investigation, that gelatine, when taken in the dissolved state, is again converted, in the body, into cellular tissue, membrane and cartilage ; that it may serve for the reproduction of such parts of these tissues as have been wasted, and for their growth.

And when the powers of nutrition in the whole body are affected by a change of the health, then, even should the power of forming blood remain the same, the organic force by which the constituents of the blood are transformed into cellular tissue and membranes must necessarily be enfeebled by sickness. In the sick man, the intensity of the vital force, its power to produce metamorphoses, must be diminished as well in the stomach as in all other parts of the body. In this condition, the uniform experience of practical physicians shews that gelatinous matters in a dissolved state exercise a most decided influence on the state of the health. Given in a form adapted for assimilation, they serve to husband the vital force, just as may be done, in the case of the stomach, by due preparation of the food in general. Brittleness in the bones of graminivorous animals is clearly owing to a weakness in those parts of the organism whose function it is to convert the constituents of the blood into cellular tissue and membrane; and if we can trust to the reports of physicians who have resided in the East, the Turkish women, in their diet of rice, and in the frequent use of enemata of strong soup, have united the conditions necessary for the formation both of cellular tissue and of fat.

PART II.

THE

METAMORPHOSIS OF TISSUES.

METAMORPHOSIS OF TISSUES.

1. THE absolute identity of composition in the chief constituents of blood and the nitrogenised compounds in vegetable food would, some years ago, have furnished a plausible reason for denying the accuracy of the chemical analyses leading to such a result. At that period, experiment had not as yet demonstrated the existence of numerous compounds, both containing nitrogen and devoid of that element, which, with the greatest diversity in external characters, yet possess the very same composition in 100 parts; nay, many of which even contain the same absolute amount of equivalents of each element. Such examples are now very frequent, and are known by the names of *isomeric* and *polymeric* compounds.

2. Cyanuric acid, for example, is a nitrogenised compound which crystallizes in beautiful transparent octahedrons, easily soluble in water and in acids, and very permanent. Cyamelide is a second body, absolutely insoluble in water and acids, white and opaque like porcelain or magnesia. Hydrated cyanic acid is a third compound, which is a liquid, more volatile than pure acetic acid, which blisters

the skin, and cannot be brought in contact with
water without being instantaneously resolved into
new products. These three substances not only
yield, on analysis, absolutely the same relative
weights of the same elements, but they may be
converted and reconverted into one another, even
in hermetically closed vessels—that is, without the
aid of any foreign matter. (See Appendix, 21.)
Again, among those substances which contain no
nitrogen, we have aldehyde, a combustible liquid
miscible with water, which boils at the temperature
of the hand, attracts oxygen from the atmosphere
with avidity, and is thereby changed into acetic acid.
This compound cannot be preserved, even in close
vessels; for after some hours or days, its consistence,
its volatility, and its power of absorbing oxygen, all
are changed. It deposits long, hard, needle-shaped
crystals, which at 212° are not volatilized, and the
supernatant liquid is no longer aldehyde. It now
boils at 140°, cannot be mixed with water, and when
cooled to a moderate degree crystallizes in a form
like ice. Nevertheless, analysis has proved, that
these three bodies, so different in their characters,
are identical in composition. (21)

3. A similar group of three occurs in the case of
albumen, fibrine, and caseine. They differ in external character, but contain exactly the same proportions of organic elements.

When animal albumen, fibrine, and caseine are
dissolved in a moderately strong solution of caustic

potash, and the solution is exposed for some time to a high temperature, these substances are decomposed. The addition of acetic acid to the solution causes, in all three, the separation of a gelatinous translucent precipitate, which has exactly the same characters and composition, from whichever of the three substances above mentioned it has been obtained.

MULDER, to whom we owe the discovery of this compound, found, by exact and careful analysis, that it contains the same organic elements, and exactly in the same proportion, as the animal matters from which it is prepared; insomuch, that if we deduct from the analysis of albumen, fibrine, and caseine, the ashes they yield, when incinerated, as well as the sulphur and phosphorus they contain, and then calculate the remainder for 100 parts, we obtain the same result as in the analysis of the precipitate above described, prepared by potash, which is free from inorganic matter. (22)

Viewed in this light, the chief constituents of the blood and the caseine of milk may be regarded as compounds of phosphates and other salts, and of sulphur and phosphorus, with a compound of carbon, nitrogen, hydrogen, and oxygen, in which the relative proportion of these elements is invariable; and this compound may be considered as the commencement and starting-point of all other animal tissues, because these are all produced from the blood.

These considerations induced Mulder to give to this product of the decomposition of albumen, &c. by potash, the name of *proteine* (from πρωτεύω, " I take the first rank "). The blood, or the constituents of the blood, are consequently compounds of this proteine with variable proportions of inorganic substances.

Mulder further ascertained, that the insoluble nitrogenised constituent of wheat flour (vegetable fibrine), when treated with potash, yields the very same product, proteine; and it has recently been proved that vegetable albumen and caseine are acted on by potash precisely as animal albumen and caseine are.

4. As far, therefore, as our researches have gone, it may be laid down as a law, founded on experience, that vegetables produce, in their organism, compounds of proteine ; and that out of these compounds of proteine the various tissues and parts of the animal body are developed by the vital force, with the aid of the oxygen of the atmosphere and of the elements of water.*

* The experiment of Tiedemann and Gmelin, who found it impossible to sustain the life of geese by means of boiled white of egg, may be easily explained, when we reflect that a graminivorous animal, especially when deprived of free motion, cannot obtain, from the transformation or waste of the tissues alone, enough of carbon for the respiratory process.   2 lbs. of albumen contain only $3\frac{1}{2}$ oz. of carbon, of which, among the last products of transformation, a fourth part is given off in the form of uric acid.

Now, although it cannot be demonstrated that proteine exists ready formed in these vegetable and animal products, and although the difference in their properties seems to indicate that their elements are not arranged in the same manner, yet the hypothesis of the pre-existence of proteine, as a point of departure in developing and comparing their properties, is exceedingly convenient. At all events, it is certain that the elements of these compounds assume the same arrangement when acted on by potash at a high temperature.

All the organic nitrogenised constituents of the body, how different soever they may be in composition, are derived from proteine. They are formed from it, by the addition or subtraction of the elements of water or of oxygen, and by resolution into two or more compounds.

5. This proposition must be received as an undeniable truth, when we reflect on the developement of the young animal in the egg of a fowl. The egg can be shewn to contain no other nitrogenised compound except albumen. The albumen of the yolk is identical with that of the white; (23) the yolk contains, besides, only a yellow fat, in which cholesterine and iron may be detected. Yet we see, in the process of incubation, during which no food and no foreign matter, except the oxygen of the air, is introduced, or can take part in the developement of the animal, that out of the albumen, feathers, claws, globules of the blood, fibrine,

membrane and cellular tissue, arteries and veins, are produced. The fat of the yolk may have contributed, to a certain extent, to the formation of the nerves and brain; but the carbon of this fat cannot have been employed to produce the organised tissues in which vitality resides, because the albumen of the white and of the yolk already contains, for the quantity of nitrogen present, exactly the proportion of carbon required for the formation of these tissues.

6. The true starting-point for all the tissues is, consequently, albumen; all nitrogenised articles of food, whether derived from the animal or from the vegetable kingdom, are converted into albumen before they can take part in the process of nutrition.

All the food consumed by an animal becomes in the stomach soluble, and capable of entering into the circulation. In the process by which this solution is effected, only one fluid, besides the oxygen of the air, takes a part; it is that which is secreted by the lining membrane of the stomach.

The most decisive experiments of physiologists have shewn that the process of chymification is independent of the vital force; that it takes place in virtue of a purely chemical action, exactly similar to those processes of decomposition or transformation which are known as putrefaction, fermentation, or decay (eremacausis).

7. When expressed in the simplest form, fer-

mentation, or putrefaction, may be described as a
process of transformation—that is, a new arrangement of the elementary particles, or atoms, of a
compound, yielding two or more new groups or
compounds, and caused by contact with other substances, the elementary particles of which are themselves in a state of transformation or decomposition.
It is a communication, or an imparting of a state of
motion, which the atoms of a body in a state of
motion are capable of producing in other bodies,
whose elementary particles are held together only
by a feeble attraction.

8. Thus the clear gastric juice contains a substance in a state of transformation, by the contact of which with those constituents of the food
which, by themselves, are insoluble in water, the
latter acquire, in virtue of a new grouping of
their atoms, the property of dissolving in that fluid.
During digestion, the gastric juice, when separated,
is found to contain a free mineral acid, the presence
of which checks all further change. That the food is
rendered soluble quite independently of the vitality
of the digestive organs has been proved by a number of the most beautiful experiments. Food, enclosed in perforated metallic tubes, so that it could
not come into contact with the stomach, was found
to disappear as rapidly, and to be as perfectly digested, as if the covering had been absent; and
fresh gastric juice, out of the body, when boiled
white of egg, or muscular fibre, were kept in

contact with it for a time at the temperature of the body, caused these substances to lose the solid form and to dissolve in the liquid.

9. It can hardly be doubted that the substance which is present in the gastric juice in a state of change is a product of the transformation of the stomach itself. No substances possess, in so high a degree as those arising from the progressive decomposition of the tissues containing gelatine or chondrine, the property of exciting a change in the arrangement of the elements of other compounds. When the lining membrane of the stomach of any animal, as, for example, that of the calf, is cleaned by continued washing with water, it produces no effect whatever, if brought into contact with a solution of sugar, with milk, or other substances. But if the same membrane be exposed for some time to the air, or dried, and then placed in contact with such substances, the sugar is changed, according to the state of decomposition of the animal matter, either into lactic acid, into mannite and mucilage, or into alcohol and carbonic acid; while milk is instantly coagulated. An ordinary animal bladder retains, when dry, all its properties unchanged; but when exposed to air and moisture, it undergoes a change not indicated by any obvious external signs. If, in this state, it be placed in a solution of sugar of milk, that substance is quickly changed into lactic acid.

10. The fresh lining membrane of the stomach of

a calf, digested with weak muriatic acid, gives to this fluid no power of dissolving boiled flesh or coagulated white of egg. But if previously allowed to dry, or if left for a time in water, it then yields, to water acidulated with muriatic acid, a substance in minute quantity, the decomposition of which is already commenced, and is completed in the solution. If coagulated albumen be placed in this solution, the state of decomposition is communicated to it, first at the edges, which become translucent, pass into a mucilage, and finally dissolve. The same change gradually affects the whole mass, and at last it is entirely dissolved, with the exception of fatty particles, which render the solution turbid. Oxygen is conveyed to every part of the body by the arterial blood; moisture is everywhere present; and thus we have united the chief conditions of all transformations in the animal body.

Thus, as in the germination of seeds, the presence of a body in a state of decomposition or transformation, which has been called *diastase*, effects the solution of the starch—that is, its conversion into sugar; so, a product of the metamorphosis of the substance of the stomach, being itself in a state of metamorphosis which is completed in the stomach, effects the dissolution of all such parts of the food as are capable of assuming a soluble form. In certain diseases, there are produced from the starch, sugar, &c., of the food, lactic acid and mucilage. (24) These are the very same products which we can produce out of sugar by means of membrane in a state of decomposition out of the body; but in a normal state of health, no lactic acid is formed in the stomach.

11. The property possessed by many substances, such as starch and the varieties of sugar, by contact with animal substances in a state of decomposition, to pass into lactic acid, has induced physiologists, without further inquiry, to assume the fact of the production of lactic acid during digestion; and the power which this acid has of dissolving phosphate of lime has led them to ascribe to it the character of a general solvent. But neither Prout nor Braconnet could detect lactic acid in the gastric juice; and even Lehmann (see his "Lehrbuch der Physiologischen Chemie," tom. i. p. 285) obtained from the gastric juice of a cat only microscopic crystals, which he took for lactate of zinc, although their chemical character could not be ascertained. The presence of free muriatic acid in the gastric juice, first observed by Prout, has been confirmed by all those chemists who have examined that fluid since. This muriatic acid is obviously derived from common salt, the soda of which plays a very decided part in the conversion of fibrine and caseine into blood.

Muriatic acid yields to no other acid in the power of dissolving bone earth, and even acetic acid, in this respect, is equal to lactic acid. There is consequently no proof of the necessity of lactic acid in the digestive process; and we know with certainty, that in artificial digestion it is not formed. Berze-

lius indeed has found lactic acid in the blood and
flesh of animals; but when his experiments were
made, chemists were ignorant of the extraordinary
facility and rapidity with which this acid is formed
from a number of substances containing its elements,
when in contact with animal matter.

In the gastric juice of a dog, Braconnet found,
along with free muriatic acid, distinct traces of a salt
of iron, which he at first held to be an accidental
admixture. But in the gastric juice of a second
dog, collected with the utmost care, the iron was
again found. (Ann. de Ch. et de Ph. lix. p. 249.)
This occurrence of iron is full of significance in
regard to the formation of the blood.

12. In the action of the gastric juice on the
food, no other element takes a share, except the
oxygen of the atmosphere and the elements of water.
This oxygen is introduced directly into the stomach.
During the mastication of the food, there is secreted
into the mouth from organs specially destined to
this function, a fluid, the saliva, which possesses the
remarkable property of enclosing air in the shape of
froth, in a far higher degree than even soap-suds.
This air, by means of the saliva, reaches the stomach
with the food, and there its oxygen enters into combination, while its nitrogen is given out through the
skin and lungs. The longer digestion continues, that
is, the greater the resistance offered to the solvent
action by the food, the more saliva, and consequently
the more air enters the stomach. Rumination, in

certain graminivorous animals, has plainly for one
object a renewed and repeated introduction of oxygen ; for a more minute mechanical division of the
food only shortens the time required for solution.

The unequal quantities of air which reach the
stomach with the saliva in different classes of animals explain the accurate observations made by
physiologists, who have established beyond all doubt
the fact, that animals give out pure nitrogen through
the skin and lungs, in variable quantity. This fact
is so much the more important, as it furnishes the
most decisive proof, that the nitrogen of the air is
applied to no use in the animal economy.

The fact that nitrogen is given out by the skin
and lungs, is explained by the property which animal
membranes possess of allowing all gases to permeate
them, a property which can be shewn to exist by the
most simple experiments. A bladder, filled with
carbonic acid, nitrogen, or hydrogen gas, if tightly
closed and suspended in the air, loses in twenty-four
hours the whole of the enclosed gas ; by a kind of
exchange, it passes outwards into the atmosphere,
while its place is occupied by atmospherical air. A
portion of intestine, a stomach, or a piece of skin
or membrane, acts precisely as the bladder, if filled
with any gas. This permeability to gases is a mechanical property, common to all animal tissues ;
and it is found in the same degree in the living as
in the dead tissue.

It is known that in cases of wounds of the lungs

a peculiar condition is produced, in which, by the
act of inspiration, not only oxygen but atmospherical air, with its whole amount (⅘ths) of nitrogen,
penetrates into the cells of the lungs. This air is
carried by the circulation to every part of the body,
so that every part is inflated or puffed up with the
air, as with water in dropsy. This state ceases,
without pain, as soon as the entrance of the air
through the wound is stopped. There can be no
doubt that the oxygen of the air, thus accumulated in the cellular tissue, enters into combination,
while its nitrogen is expired through the skin and
lungs.

Moreover, it is well known that in many graminivorous animals, when the digestive organs have been
overloaded with fresh juicy vegetables, these substances undergo in the stomach the same decomposition as they would at the same temperature out of
the body. They pass into fermentation and putrefaction, whereby so great a quantity of carbonic acid
gas and of inflammable gas is generated, that these
organs are enormously distended, sometimes even to
bursting. From the structure of their stomach or
stomachs, these gases cannot escape through the
œsophagus; but in the course of a few hours, the
distended body of the animal becomes less swoln,
and at the end of twenty-four hours no trace of the
gases is left. (25)

Finally, if we consider the fatal accidents which
so frequently occur in wine countries from the

drinking of what is called feather-white wine (*der federweisse Wein*), we can no longer doubt that gases of every kind, whether soluble or insoluble in water, possess the property of permeating animal tissues, as water penetrates unsized paper. This poisonous wine is wine still in a state of fermentation, which is increased by the heat of the stomach. The carbonic acid gas which is disengaged penetrates through the parietes of the stomach, through the diaphragm, and through all the intervening membranes, into the air-cells of the lungs, out of which it displaces the atmospherical air. The patient dies with all the symptoms of asphyxia caused by an irrespirable gas ; and the surest proof of the presence of the carbonic acid in the lungs is the fact, that the inhalation of ammonia (which combines with it) is recognized as the best antidote against this kind of poisoning.

The carbonic acid of effervescing wines and of soda-water, when taken into the stomach, or of water saturated with this gas, administered in the form of enema, is given out again through the skin and lungs ; and this is equally true of the nitrogen which is introduced into the stomach with the food in the saliva.

No doubt a part of these gases may enter the venous circulation through the absorbent and lymphatic vessels, and thus reach the lungs, where they are exhaled ; but the presence of membranes offers not the slightest obstacle to their passing directly into the cavity of the chest. It is, in fact, difficult to suppose that the absorbents and lymphatics have any peculiar tendency to absorb air, nitrogen, or hydrogen, and convey these gases into the circulation, since the intestines, the stomach, and all spaces in the body not filled with solid or liquid matters, contain gases, which only quit their position when their volume exceeds a certain point, and which, consequently, are not absorbed. More especially in reference to nitrogen, we must suppose that it is removed from the stomach by some more direct means, and not by the blood, which fluid must already, in passing through the lungs, have become saturated with that gas, hat is, must have absorbed a quantity of it, proportioned to its solvent power, like any other liquid. By the respiratory motions all the gases which fill the otherwise empty spaces of the body are urged towards the chest; for by the motion of the diaphragm and the expansion of the chest a partial vacuum is produced, in consequence of which air is forced into the chest from all sides by the atmospheric pressure. The equilibrium is, no doubt, restored, for the most part, through the windpipe, but all the gases in the body must, nevertheless, receive an impulse towards the chest. In birds and tortoises these arrangements are reversed. If we assume that a man introduces into the stomach in each minute only $\frac{1}{8}$th of a cubic inch of air with the saliva, this makes in eighteen hours 135 cubic inches; and if $\frac{1}{5}$th be deducted as oxygen, there will still remain 108 cubic inches of nitrogen, which occupy the space of 3 lbs. of water. Now whatever may be the actual amount of the nitrogen thus swallowed, it is certain that the whole of it is given out again by the mouth, nose, and skin; and when we consider the very large quantity of nitrogen found in the intestines of executed criminals by Magendie, as well as the entire absence of oxygen in these organs (26), we must assume that air, and consequently nitrogen, enters the stomach by resorption through the skin, and is afterwards exhaled by the lungs.

When animals are made to respire in gases containing no nitrogen, more of that gas is exhaled, because in this case the nitrogen within the body acts towards the external space as if the latter were a vacuum. (See Graham, " On the Diffusion of Gases.")

The differences in the amount of expired nitrogen in different classes of animals are thus easily explained ; the herbivora swallow with the saliva more air than the carnivora; they expire more nitrogen than the latter,—less when fasting than immediately after taking food.

13. In the same way as muscular fibre, when separated from the body, communicates the state of decomposition existing in its elements to the peroxide of hydrogen, so a certain product, arising by means of the vital process, and in consequence of the transposition of the elements of parts of the stomach and of the other digestive organs, while its own

metamorphosis is accomplished in the stomach, acts on the food. The insoluble matters become soluble —they are digested.

It is certainly remarkable, that hard-boiled white of egg or fibrine, when rendered soluble by certain liquids, by organic acids, or weak alkaline solutions, retain all their properties except the solid form (cohesion) without the slightest change. Their elementary molecules, without. doubt, assume a new arrangement ; they do not, however, separate into two or more groups, but remain united together.

The very same thing occurs in the digestive process ; in the normal state, the food only undergoes a change in its state of cohesion, becoming fluid without any other change of properties.

The greatest obstacle to forming a clear conception of the nature of the digestive process, which is here reckoned among those chemical metamorphoses which have been called fermentation and putrefaction, consists in our involuntary recollection of the phenomena which accompany the fermentation of sugar and of animal substances (putrefaction), which phenomena we naturally associate with any similar change ; but there are numberless cases in which a complete chemical metamorphosis of the elements of a compound occurs without the smallest disengagement of gas, and it is chiefly these which must be borne in mind, if we would acquire a clear and accurate idea of the chemical notion or conception of the digestive process.

All substances which can arrest the phenomena of fermentation and putrefaction in liquids, also arrest digestion when taken into the stomach. The action of the empyreumatic matters in coffee and tobacco-smoke, of creosote, of mercurials, &c. &c., is on this account worthy of peculiar attention with reference to dietetics.

The identity in composition of the chief constituents of blood and of the nitrogenised constituents of vegetable food has certainly furnished, in an unexpected manner, an explanation of the fact that putrefying blood, white of egg, flesh, and cheese produce the same effects in a solution of sugar as yeast or ferment ; that sugar, in contact with these substances, according to the particular stage of decomposition in which the putrefying matters may be, yields, at one time, alcohol and carbonic acid ; at another, lactic acid, mannite, and mucilage. The explanation is simply this, that ferment, or yeast, is nothing but vegetable fibrine, albumen, or caseine in a state of decomposition, these substances having the same composition with the constituents of flesh, blood, or cheese. The putrefaction of these animal matters is a process identical with the metamorphosis of the vegetable matters identical with them ; it is a separation or splitting up into new and less complex compounds. And if we consider the transformation of the elements of the animal body (the waste of matter in animals) as a chemical process which goes on under the influence of the vital force, then the putrefaction of animal matters out of the body is a division into simpler compounds, in which the vital force takes no share. The action in both cases is the same, only the products differ. The practice of medicine has furnished the most beautiful and interesting observations on the action of empyreumatic substances, such as wood, vinegar, creosote, &c., on malignant wounds and ulcers. In such morbid phenomena two actions are going on together; one metamorphosis, which strives to complete itself under the influence of the vital force, and another, independent of that force. The latter is a chemical process, which is entirely suppressed or arrested by empyreumatic substances; and this effect is precisely opposed to the poisonous influence exercised on the organism by putrefying blood when introduced into a fresh wound.

14. The formula $C_{48}H_{36}N_6O_{14}$* is that which most accurately expresses the composition of proteine, or the relative proportions of the organic elements in the blood, as ascertained by analysis. Albumen, fibrine, and caseine contain proteine; caseine contains, besides, sulphur, but no phosphorus; albumen and fibrine contain both these substances chemically combined—the former more sulphur than the latter. We cannot directly ascertain in what form the phosphorus exists, but we have decided proof that the sulphur cannot be in the

* For the method of converting this and other formulæ into proportions per cent. see Appendix.

oxidised state. All these substances, when heated
with a moderately strong solution of potash, yield
the sulphur which we find in the solution as sulphuret of potassium ; and on the addition of an
acid it is given off as sulphuretted hydrogen. When
pure fibrine or ordinary albumen is dissolved in a
weak solution of potash, and acetate of lead is added
to the solution, in such proportion that the whole of
the oxide of lead remains dissolved in the potash,
the mixture, if heated to the boiling point, becomes
black like ink, and sulphuret of lead is deposited
as a fine black powder.

It is extremely probable, that by the action of
the alkali the sulphur is removed as sulphuretted
hydrogen, the phosphorus as phosphoric or phos-
phorous acid. Since, in this case, sulphur and phosphorus are eliminated on the one hand, and oxygen
and hydrogen on the other, it might be concluded
that fibrine and albumen, when analysed with their
sulphur and phosphorus, would yield a larger proportion of oxygen and hydrogen than is found in
proteine. But this cannot be shewn in the analysis;
for fibrine, for example, has been found to contain
0·36 per cent. of sulphur. Assuming, then, that
this sulphur is eliminated by the alkali in combination with hydrogen, proteine would yield 0·0225
per cent. less hydrogen than fibrine; instead of the
mean amount of 7·062 per cent. of hydrogen, the
proteine should yield 7·04 per cent. In like manner, by the elimination of the phosphorus in combi-

nation with oxygen, the amount of oxygen in fibrine would be reduced from 22·715—22·00 per cent. to 22·5—21·8 per cent. in proteine. But the limits of error in our analyses are, on an average, beyond $\frac{1}{10}$th per cent. in the hydrogen, and beyond $\frac{1}{10}$ths per cent. in the oxygen; while in the supposed case the difference in the hydrogen would not be greater than $\frac{1}{48}$th per cent.

Finally, if we reflect, that the elimination of oxygen and hydrogen with the sulphur and phosphorus does not exclude the addition of the elements of water, and if we assume that fibrine and albumen, in passing into proteine, do combine with a certain quantity of water, an occurrence which is highly probable, we shall see that there is no probability that the ultimate analysis of these compounds shall ever enable us to decide such questions, or to fix the chemical view of the relation of proteine to albumen, fibrine, or caseine, farther than has been done above.

Some have endeavoured to prove the existence of unoxidised phosphorus in albumen and fibrine from the formation of sulphuret of potassium when they are acted on by potash, supposing the oxygen of the potash to have formed phosphoric acid with the phosphorus; but caseine, which contains no phosphorus, yields sulphuret of potassium, just like the other substances; and here its formation cannot be accounted for, unless we admit the previous production of sulphuretted hydrogen. In the mere boiling of flesh, for the purpose of making soup, sulphuretted hydrogen, as Chevreul has shewn, is disengaged.

Moreover, the proportion of sulphur, for the same amount of phosphorus, is not the same in fibrine and albumen, from which no other conclusion can be drawn, but that the formation of sulphuret of potassium has no relation to the presence of phosphorus. Sulphuret of potassium is formed from caseine, which is not supposed to contain any uncombined phosphorus; and it is formed, also, from albumen, which contains only half as much phosphorus as fibrine.

Every attempt to give the true absolute amount of the atoms in fibrine and albumen in a rational formula, in which the sulphur and phosphorus are taken, not in fractions, but in entire equivalents, must be fruitless, because we are absolutely unable to determine with perfect accuracy the exceedingly minute quantities of sulphur and phosphorus in such compounds; and because a variation in the sulphur or phosphorus, smaller in extent than the usual limit of errors of observation, will affect the number of atoms of carbon, hydrogen, or oxygen to the extent of 10 atoms or more.

We must be careful not to deceive ourselves in our expectations of what chemical analysis can do. We know, with certainty, that the numbers representing the relative proportions of the organic elements are the same in albumen and fibrine, and hence we conclude that they have the same composition. This conclusion is not affected by the fact, that we do not know the absolute number of the atoms of their elements, which have united to form the compound atom.

15. A formula for proteine is nothing more than the nearest and most exact expression in equivalents, of the result of the best analyses; it is a fact established so far, free from doubt, and this alone is, for the present, valuable to us.

If we reflect, that from the albumen and fibrine of the body all the other tissues are derived, it is perfectly clear, that this can only occur in two ways. Either certain elements have been added to, or removed from, their constituent parts.

If we now, for example, look for an analytical expression of the composition of cellular tissue, of the tissues yielding gelatine, of tendons, of hair, of horn, &c., in which the number of atoms of carbon is made invariably the same as in albumen and fibrine, we can then see, at the first glance, in what way the proportion of the other elements has been altered; but this includes all that physiology requires in order to obtain an insight into the true nature of the formative and nutritive processes in the animal body.

From the researches of Mulder and Scherer we obtain the following empirical formulæ :—

Composition of organic tissues.

Albumen .... $C_{48}N_6H_{36}O_{14} + P + S*$
Fibrine .... $C_{48}N_6H_{36}O_{14} + P + 2S$
Caseine .... $C_{48}N_6H_{36}O_{14} + S$
Gelatinous tissues, tendons ... $C_{48}N_{7.5}H_{41}O_{18}$
Chondrine .... $C_{48}N_6H_{40}O_{20}$
Hair, horn .... $C_{48}N_7H_{39}O_{17}$
Arterial membrane‧ .... $C_{48}N_6H_{38}O_{16}$

The composition of these formulæ shews, that when proteine passes into chondrine (the substance of the cartilages of the ribs), the elements of water, with oxygen, have been added to it ; while in the formation of the serous membranes, nitrogen also has entered into combination.

If we represent the formula of proteine, $C_{48}N_6H_{36}O_{14}$ by Pr, then nitrogen, hydrogen, and oxygen have been added to it in the form of known compounds, and in the following proportions, in forming the gelatinous tissues, hair, horn, arterial membrane, &c.

	Proteine.	Ammonia.	Water.	Oxygen.
Fibrine, Albumen......	Pr			
Arterial membrane ...	Pr		+ 2HO	
Chondrine	Pr		+ 4HO	+ 2O.
Hair, horn	Pr +	NH_3		+ 3O.
Gelatinous tissues ...	2Pr +	$3NH_3$ +	HO	+ 7O.

17. From this general statement it appears that all the tissues of the body contain, for the same

* The quantities of sulphur and phosphorus here expressed by S and P are not equivalents, but only give the relative proportions of these two elements to each other, as found by analysis.

amount of carbon, more oxygen than the constituents of blood. During their formation, oxygen, either from the atmosphere or from the elements of water, has been added to the elements of proteine. In hair and gelatinous membrane we observe, further, an excess of nitrogen and hydrogen, and that in the proportions to form ammonia.

Chemists are not yet agreed on the question, in what manner the elements of sulphate of potash are arranged; it would therefore be going too far, were they to pronounce arterial membrane a hydrate of proteine, chondrine a hydrated oxide of proteine, and hair and membranes compounds of ammonia with oxides of proteine.

The above formulæ express with precision the differences of composition in the chief constituents of the animal body; they shew, that for the same amount of carbon the proportion of the other elements varies, and how much more oxygen or nitrogen one compound contains than another.

18. By means of these formulæ we can trace the production of the different compounds from the constituents of blood; but the explanation of their production may take two forms, and we have to decide which of these comes nearest to the truth.

For the same amount of carbon, membranes and the tissues which yield gelatine contain more nitrogen, oxygen, and hydrogen than proteine. It is conceivable that they are formed from albumen by the addition of oxygen, of the elements of water, and of those of ammonia, accompanied by the separation of sulphur and phosphorus; at all events, their composition is entirely different from that of the chief constituents of blood.

The action of caustic alkalies on the tissues yielding gelatine shews distinctly that they no longer contain proteine; that substance cannot in any way be obtained from them; and all the products formed by the action of alkalies on them differ entirely from those produced by the compounds of proteine in the same circumstances. Whether proteine exist, ready formed, in fibrine, albumen, and caseine, or not, it is certain that their elements, under the influence of the alkali, arrange themselves so as to form proteine; but this property is wanting in the elements of the tissues which yield gelatine.

The other, and perhaps the more probable explanation of the production of these tissues from proteine, is that which makes it dependent on a separation of carbon.

If we assume the nitrogen of proteine to remain entire in the gelatinous tissue, then the composition of the latter, calculated on 6 equivalents of nitrogen, would be represented by the formula, $C_{38}N_6H_{32}O_{14}$. This formula approaches most closely to the analysis of Scherer, although it is not an exact expression of his results. A formula corresponding more perfectly to the analyses, is $C_{32}N_5H_{27}O_{12}$; or, calculated according to Mulder's analysis, $C_{54}N_9H_{42}O_{20}$.*

* The formula $C_{52}N_8H_{40}O_{20}$, adopted by Mulder, gives, when

According to the first formula, carbon and hydrogen have been separated; according to the two last, a certain proportion of all the elements has been removed.

19. We must admit, as the most important result of the study of the composition of gelatinous tissue, and as a point undeniably established, that, although formed from compounds of proteine, it no longer belongs to the series of the compounds of proteine. Its chemical characters and composition justify this conclusion.

No fact is as yet opposed to the law, deduced from observation, that nature has exclusively destined compounds of proteine for the production of blood.

No substance analogous to the tissues yielding gelatine is found in vegetables. The gelatinous substance is not a compound of proteine; it contains no sulphur, no phosphorus, and it contains more nitrogen or less carbon than proteine. The compounds of proteine, under the influence of the vital energy of the organs which form the blood, assume a new form, but are not altered in composition; while these organs, as far as our experience reaches, do not possess the power of producing compounds of proteine, by virtue of any influence, out of substances which contain no proteine. Animals which were fed exclusively with gelatine, the

reduced to 100 parts, too little nitrogen to be considered an exact expression of his analyses.

most highly nitrogenised element of the food of carnivora, died with the symptoms of starvation ; in short, the gelatinous tissues are incapable of conversion into blood.

But there is no doubt that these tissues are formed from the constituents of the blood ; and we can hardly avoid entertaining the supposition, that the fibrine of venous blood, in becoming arterial fibrine, passes through the first stage of conversion into gelatinous tissue.   We cannot, with much probability, ascribe to membranes and tendons the power of forming themselves out of matters brought by the blood ; for how could any matter become a portion of cellular tissue, for example, by virtue of a force which has as yet no organ ?   An already existing cell may possess the power of reproducing or of multiplying itself, but in both cases the presence of a substance identical in composition with cellular tissue is essential.   Such matters are formed in the organism, and nothing can be better fitted for their production than the substance of the cells and membranes which exist in animal food, and become soluble in the stomach during digestion, or which are taken by man in a soluble form.

20. In the following pages I offer to the reader an attempt to develope analytically the principal metamorphoses which occur in the animal body ; and, to preclude all misapprehension, I do this with a distinct protest against all conclusions and deductions which may now or at any subsequent period be

derived from it in opposition to the views developed in the preceding part of this work, with which it has no manner of connection. The results here to be described have surprised me no less than they will others, and have excited in my mind the same doubts as others will conceive ; but they are not the creations of fancy, and I give them because I entertain the deep conviction that the method which has led to them is the only one by which we can hope to acquire insight into the nature of the organic processes.

The numberless qualitative investigations of animal matters which are made are equally worthless for physiology and for chemistry, so long as they are not instituted with a well-defined object, or to answer a question clearly put.

If we take the letters of a sentence which we wish to decipher, and place them in a line, we advance not a step towards the discovery of their meaning. To resolve an enigma, we must have a perfectly clear conception of the problem. There are many ways to the highest pinnacle of a mountain ; but those only can hope to reach it who keep the summit constantly in view. All our labour and all our efforts, if we strive to attain it through a morass, only serve to cover us more completely with mud ; our progress is impeded by difficulties of our own creation, and at last even the greatest strength must give way when so absurdly wasted.

21. If it be true that all parts of the body are

formed and developed from the blood or the constituents of the blood, that the existing organs at every moment of life are transformed into new compounds under the influence of the oxygen introduced in the blood, then the animal secretions must of necessity contain the products of the metamorphosis of the tissues.

22. If it be further true, that the urine contains those products of metamorphosis which contain the most nitrogen, and the bile those which are richest in carbon, from all the tissues which in the vital process have been transformed into unorganised compounds, it is clear that the elements of the bile and of the urine, added together, must be equal, in the relative proportion of these elements to the composition of the blood.

23. The organs are formed from the blood, and contain the elements of the blood ; they become transformed into new compounds, with the addition only of oxygen and of water. Hence the relative proportion of carbon and nitrogen must be the same as in the blood.

If then we subtract from the composition of blood the elements of the urine, then the remainder, deducting the oxygen and water which have been added, must give the composition of the bile.

Or if from the elements of the blood, we subtract the elements of the bile, the remainder must give the composition of urate of ammonia, or of urea and carbonic acid.

It will surely appear remarkable that this manner of viewing the subject has led to the true formula of bile, or, to speak more accurately, to the most correct empirical expression of its composition; and has furnished the key to its metamorphoses, under the influence of acids and alkalies, which had previously been sought for in vain.

24. When fresh drawn blood is made to trickle over a plate of silver, heated to 140°, it dries to a red, varnish-like matter, easily reduced to powder. Muscular flesh, free from fat, if dried first in a gentle heat, and then at 212°, yields a brown, pulverizable mass.

The analyses of PLAYFAIR and BOECKMANN (28) give for flesh (fibrine, albumen, cellular tissue, and nerves) and for blood, as the most exact expression of their numerical results, one and the same formula, namely, $C_{48}N_6H_{39}O_{15}$. This may be called the empirical formula of blood.

25. The chief constituent of bile, according to the researches of DEMARÇAY, is a compound, analogous to soaps, of soda with a peculiar substance, which has been named *choleic acid*. This acid is obtained in combination with oxide of lead, when bile, purified by means of alcohol from all matters insoluble in that menstruum, is mixed with acetate of lead.

Choleic acid is resolved, by the action of muriatic acid, into *ammonia, taurine,* and a new acid, *choloidic acid*, which contains no nitrogen.

When boiled with caustic potash, choleic acid
is resolved into *carbonic acid*, *ammonia*, and another
new acid, *cholic acid* (distinct from the cholic acid
of Gmelin).

Now it is clear that the true formula of choleic
acid must include the analytical expression of these
modes of decomposition; in other words, that it
must enable us to shew that the composition of the
products derived from it is related, in a clear and
simple manner, to the composition of the acid itself.
This is the only satisfactory test of a formula; and
the analytical expression thus obtained loses nothing
of its truth or value, if it should appear, as the researches of BERZELIUS seem to shew, that choleic
and choloidic acids are mixtures of different compounds; for the relative proportions of the elements cannot in any way be altered by this circumstance.

26. In order to develope the metamorphoses
which choleic acid suffers under the influence of
acids and alkalies, the following formula alone can
be adopted as the empirical expression of the results
of its analysis.

Formula of choleic acid : $C_{76}N_2H_{66}O_{22}$. (29)

I repeat, that this formula may express the composition of one, or of two or more compounds;
no matter of how many compounds the so-called
choleic acid may be made up, the above formula
represents the relative proportions of all their elements taken together.

If now we subtract from the elements of choleic acid, the products formed by the action of muriatic acid, namely, ammonia and taurine, we obtain the empirical formula of choloidic acid. Thus: from the

Formula of choleic acid .... $C_{76}N_2H_{66}O_{22}$
   Subtract—

1 atom taurine......... $C_4NH_7O_{10}$ ⎱ $= C_4\,N_2H_{10}O_{10}$
1 eq. ammonia......... NH_3 ⎰

There remains the formula of cho-
   loidic acid .... $= C_{72}\quad H_{56}O_{12}$ (30)

27. Again, if from the formula of choleic acid we subtract the elements of urea and 2 atoms of water (= 2 eq. carbonic acid and 2 eq. ammonia), there will remain the formula and composition of cholic acid. Thus; from the

Formula of choleic acid .... $= C_{76}N_2H_{66}O_{22}$
   subtract—

2 eq. carbonic acid $= C_2\qquad O_4$ ⎱ $= C_2\,N_2H_6\,O_4$
2 eq. ammonia $=\quad N_2H_6$ ⎰

Remains the formula of cholic acid $= C_{74}\quad H_{60}O_{18}$ (31)

When we consider the very close coincidence between these formulæ and the actual results of analysis (see Appendix, 29, 30, 31), it is scarcely possible to doubt that the formula above adopted for choleic acid expresses, as accurately as is to be expected in the analysis of such compounds, the relative proportion of its elements, no matter in how many different forms they may be united to produce that acid.

28. Let us now add the half of the numbers which represent the formula of choleic acid, to the

elements of the urine of serpents—that is, to neutral urate of ammonia, as follows :

$\frac{1}{2}$ the formula of choleic acid...... $= C_{38}N\,H_{33}O_{11}$
Add to this—
1 eq. uric acid...... $= C_{10}N_4H_4O_6$ $\left.\vphantom{\begin{matrix}a\\b\end{matrix}}\right\} = C_{10}N_5H_7\,O_6$
1 eq. ammonia...... $=\quad N\,H_3$
The sum is .... $= \overline{C_{48}N_6H_{40}O_{17}}$

29. But this last formula expresses the composition of blood, with the addition of 1 eq. oxygen and 1 eq. water.

Formula of blood .... $C_{48}N_6H_{39}O_{15}$
1 eq. water .... $= HO$ $\left.\vphantom{\begin{matrix}a\\b\end{matrix}}\right\} =\quad H\,O_2$
1 eq. oxygen................... $=\quad O$
The sum is........................... $= \overline{C_{48}N_6H_{40}O_{17}}$

30. If, moreover, we add to the elements of proteine those of 3 eq. water, we obtain, with the exception of 1 eq. hydrogen, exactly the same formula.

Formula of proteine................... $= C_{48}N_6H_{36}O_{14}$
Add 3 eq. of water .... $=\quad H_3\,O_3$
The sum is .... $\overline{C_{48}N_6H_{39}O_{17}}$

differing only by 1 eq. of hydrogen from the formula above obtained by adding together choleic acid and urate of ammonia.

31. If, then, we consider choleic acid and urate of ammonia the products of the transformation of muscular fibre, since no other tissue in the body contains proteine (for albumen passes into tissues, without our being able to say, that in the vital process it is directly resolved into choleic acid, and urate of ammonia), there exist in fibrine, with the

addition of the elements of water, all the elements essential to this metamorphosis ; and, except the sulphur and phosphorus, both of which are probably oxidised, no element is separated.

This form of metamorphosis is applicable to the vital transformations in the lower classes of amphibia, and perhaps in worms and insects. In the higher classes of animals the uric acid disappears in the urine, and is replaced by urea.

The disappearance of uric acid and the production of urea plainly stand in a very close relation to the amount of oxygen absorbed in respiration, and to the quantity of water consumed by different animals in a given time.

When uric acid is subjected to the action of oxygen, it is first resolved, as is well known, into alloxan and urea. (32) A new supply of oxygen acting on the alloxan causes it to resolve itself either into oxalic acid and urea, into oxaluric and parabanic acids, (33) or into carbonic acid and urea.

32. In the so-called mulberry calculi we find oxalate of lime, in other calculi urate of ammonia, and always in persons, in whom, from want of exercise and labour, or from other causes, the supply of oxygen has been diminished. Calculi containing uric acid or oxalic acid are never found in phthisical patients ; and it is a common occurrence in France, among patients suffering from calculous complaints, that when they go to the country, where they take more exercise, the compounds of uric acid, which

were deposited in the bladder during their residence in town, are succeeded by oxalates (mulberry calculus), in consequence of the increased supply of oxygen. With a still greater supply of oxygen they would have yielded, in healthy subjects, only the last product of the oxidation of uric acid, namely, carbonic acid and urea.

An erroneous interpretation of the undeniable fact that all substances incapable of further use in the organism are separated by the kidneys and expelled from the body in the urine, altered or unaltered, has led practical medical men to the idea, that the food, and especially nitrogenised food, may have a direct influence on the formation of urinary calculi. There are no reasons which support this opinion, while those opposed to it are innumerable. It is possible that there may be taken, in the food, a number of matters changed by the culinary art, which, as being no longer adapted to the formation of blood, are expelled in the urine, more or less altered by the respiratory process. But roasting and boiling alter in no way the composition of animal food. (34)

Boiled and roasted flesh is converted at once into blood; while the uric acid and urea are derived from the metamorphosed tissues. The quantity of these products increases with the rapidity of transformation in a given time, but bears no proportion to the amount of food taken in the same period. In a starving man who is any way compelled to undergo severe and continued exertion, more urea is secreted than in the most highly fed individual, if in a state of rest. In fevers and during rapid emaciation the urine contains more urea than in the state of health. (PROUT.)

33. In the same way, therefore, as the hippuric acid, present in the urine of the horse when at rest, is converted into benzoate of ammonia and carbonic acid as soon as the animal is compelled to labour, so the uric acid disappears in the urine of man, when he receives, through the skin and lungs, a quantity of oxygen sufficient to oxidise the products of the transformation of the tissues. The use of wine and fat, which are only so far altered in the organism that they combine with oxygen, has a marked influence on the formation of uric acid. The urine, after fat food has been taken, is turbid, and deposits minute crystals of uric acid. (PROUT.) The same thing is observed after the use of wines in which the alkali necessary to retain the uric acid in solution is wanting, but never from the use of Rhenish wines, which contain so much tartar.

In animals which drink much water, by means of which the sparingly soluble uric acid is kept dissolved, so that the inspired oxygen can act on it, no uric acid is found in the urine, but only urea. In birds, which seldom drink, uric acid predominates.

If to 1 atom of uric acid we add 6 atoms of oxygen and 4 atoms of water, it resolves itself into urea and carbonic acid:

$$\left.\begin{array}{l}\text{1 at. uric acid } C_{10}N_4H_4O_6 \\ \text{4 at. water} \\ \text{6 at. oxygen} \end{array}\right\} \quad H_4O_{10}\left.\begin{array}{l}\\ \\ \end{array}\right\} = \left\{\begin{array}{l} \text{2 at. urea } \ldots\ldots\ldots\ C_4\,N_4H_8O_4 \\ \text{6 at. carbonic acid } C_6 \qquad O_{12} \end{array}\right.$$

$$\overline{\qquad C_{10}N_4H_8O_{16} \qquad} \qquad\qquad \overline{C_{10}N_4H_8O_{16}}$$

34. The urine of the herbivora contains no uric acid, but ammonia, urea, and hippuric or benzoic acid. By the addition of 9 atoms of oxygen to the empirical formula of their blood multiplied by 5, we obtain the elements of 6 at. of hippuric acid, 9 at. of urea, 3 at. of choleic acid, 3 at. of water, and 3 at. of ammonia; or, if we suppose 45 atoms of oxygen to be added to the blood during its metamorphosis, then we obtain 6 at. of benzoic acid, $13\frac{1}{2}$ at. of urea, 3 at. of choleic acid, 15 at. of carbonic acid, and 12 at. of water.

$$5\,(C_{48}N_6H_{39}O_{15}) + O_9 = C_{240}N_{30}H_{195}O_{84}$$

$$=\left\{\begin{array}{lll} \text{6 at. hippuric acid,} & 6\,(C_{18}N\,H_8\,O_5\,) = & C_{108}N_6\,H_{48}\,O_{30} \\ \text{9 at. urea } \ldots\ldots\ldots & 9\,(C_2\,N_2H_4\,O_2\,) = & C_{18}\,N_{18}H_{36}\,O_{18} \\ \text{3 at. choleic acid,} & 3\,(C_{38}N\,H_{33}O_{11}) = & C_{114}N_3\,H_{99}\,O_{33} \\ \text{3 at. ammonia } \ldots & 3\,(\quad N\,H_3\quad) = & N_3\,H_9 \\ \text{3 at. water} \ldots\ldots\ldots & 3\,(\quad H_3\,O_3\,) = & H_3\ O_3 \end{array}\right.$$

The sum is $\ldots\ldots\ldots\ldots\ldots\ldots\ldots$ $C_{240}N_{30}H_{195}O_{84}$

or—

$$5\,(C_{48}N_6H_{39}O_{15}) + O_{45} = C_{240}N_{30}H_{195}O_{120}$$

$$=\left\{\begin{array}{lll} \text{6 at. benzoic acid} & 6\,(C_{14}\ H_5\ O_3\,) = & C_{84}\qquad H_{30}\,O_{18} \\ \text{27/2 at. urea} \ldots\ldots & 27\,(C\ NH_2\ O\ ) = & C_{27}\,N_{27}H_{54}\,O_{27} \\ \text{3 at. choleic acid} & 3\,(C_{38}NH_{33}O_{11}) = & C_{114}N_3\,H_{99}\,O_{33} \\ \text{15 at. carbonic acid,} & 15\,(C\qquad O_2\,) = & C_{15}\qquad O_{30} \\ \text{12 at. water} \ldots\ldots & 12\,(\quad H\ O\ ) = & H_{12}\,O_{12} \end{array}\right.$$

The sum is $\ldots\ldots\ldots\ldots\ldots\ldots\ldots$ $C_{240}N_{30}H_{195}O_{120}$

35. Lastly, let us follow the metamorphosis of

the tissues in the fœtal calf, considering the proteine furnished in the blood of the mother as the substance which undergoes or has undergone a transformation; it will appear that 2 at. of proteine, without the addition of oxygen or any other foreign element, except 2 at. of water, contain the elements of 6 at. of allantoine and 1 at. of choloidic acid. (meconium ?)

$$2 \text{ at. proteine} = 2 \, (C_{48}N_6H_{36}O_{14}) + 2 \text{ at. water} = 2 \, HO = C_{96}N_{12}H_{74}O_{30}$$

$$= \begin{cases} 6 \text{ at. allantoine, } 6 \, (C_4N_2H_3O_3) &= C_{24}N_{12}H_{18}O_{18} \\ 1 \text{ at. choloidic acid} &= C_{72} \quad H_{56}O_{12} \end{cases}$$

$$\overline{C_{96}N_{12}H_{74}O_{30}}$$

36. But the elements of the six atoms of allantoine in the last equation correspond exactly to the elements of 2 at. of uric acid, 2 at. of urea, and 2 at. of water.

$$6 \text{ at. of allantoine} = C_{24}N_{12}H_{18}O_{18} = \begin{cases} 2 \text{ at. uric acid } C_{20}N_8 \, H_8 \, O_{12} \\ 2 \text{ at. urea} \quad C_4 \, N_4 \, H_8 \, O_4 \\ 2 \text{ at. water} \quad H_2 \, O_2 \end{cases}$$

$$\overline{C_{24}N_{12}H_{18}O_{18}}$$

The relations of allantoine, which is found in the urine of the fœtal calf, to the nitrogenised constituents of the urine in animals which respire, are, as may be seen by comparing the above formulæ, such as cannot be overlooked or doubted. Allantoine contains the elements of uric acid and urea—that is, of the nitrogenised products of the transformation of the compounds of proteine.

37. Further, if to the formula of proteine, multiplied by 3, we add the elements of 4 at. of water,

and if we deduct from the sum of all the elements half of the elements of choloidic acid, there remains a formula which expresses very nearly the composition of gelatine. From

$$3 \, (C_{48}N_6H_{36}O_{14}) + 4 \, HO\ldots = C_{144}N_{18}H_{112}O_{46}$$
$$\text{Subtract } \tfrac{1}{2} \text{ at. choloidic acid} = C_{36} \quad H_{28} \, O_6$$

There remain .... $C_{108}N_{18}H_{84}O_{40}$ (35)

38. Subtracting from this formula of gelatine the elements of 2 at. of proteine, there remain the elements of urea, uric acid, and water, or of 3 at. of allantoine and 3 at. of water. Thus—

$$\text{Formula of gelatine (Mulder) } C_{108}N_{18}H_{84}O_{40}$$
$$\text{Subtract 2 at. proteine } \ldots \ldots C_{96}N_{12}H_{72}O_{28}$$

There remain............ $C_{12}N_6H_{12}O_{12} =$

$$=\begin{cases} 1 \text{ at. uric acid } C_{10}N_4H_4 \, O_6 \\ 1 \text{ at. urea } \ldots C_2 \, N_2H_4 \, O_2 \\ 4 \text{ at. water } \qquad H_4 \, O_4 \end{cases} = \begin{cases} 3 \text{ at. allantoine } \quad C_{12}N_6H_9 \, O_9 \\ 3 \text{ at. water } \ldots \qquad H_3 \, O_3 \end{cases}$$

$$\overline{C_{12}N_6H_{12}O_{12}} \qquad\qquad \overline{C_{12}N_6H_{12}O_{12}}$$

39. The numerical proportions calculated from the above formula differ from those actually obtained in the analyses of MULDER and SCHERER in this, that the latter indicate somewhat less of nitrogen in gelatine; but if we assume the formula to be correct, it then appears, from the statement just given, that the elements of two atoms of proteine, *plus* the nitrogenised products of the transformation of a third atom of proteine (uric acid and urea) and water; or three atoms of proteine, *minus* the elements of a compound containing no nitrogen, which

actually occurs as one of the products of the transformation of choleic acid, yield in both cases a formula closely approaching to the composition of gelatinous tissues. We must, however, attach to such formulæ, and to the considerations arising from them, no more importance than justly belongs to them. I would constantly remind the reader that their use is to serve as points of connection, which may enable us to acquire more accurate views as to the production and decomposition of those compounds which form the animal tissues. They are the first attempts to discover the path which we must follow in order to attain the object of our researches; and this object, the goal we strive to reach, is, and must be, attainable.

The experience of all those who have occupied themselves with researches into natural phenomena leads to this general result, that these phenomena are caused, or produced, by means far more simple than was previously supposed, or than we even now imagine; and it is precisely their simplicity which should most powerfully excite our wonder and admiration.

Gelatinous tissue is formed from blood, from compounds of proteine. It may be produced by the addition, to the elements of proteine, of allantoine and water, or of water, urea, and uric acid; or by the separation from the elements of proteine of a compound containing no nitrogen. The solution of such problems becomes less difficult, when the problem to be solved, the question to be answered, is matured and clearly put. Every experimental decision of any such question in the negative forms the starting-point of a new question, the solution of which, when obtained, is the necessary consequence of our having put the first question.

40. In the foregoing sections, no other constituent of the bile, besides choleic acid, has been brought into the calculation; because it alone is known with certainty to contain nitrogen. Now, if it be admitted that its nitrogen is derived from the metamorphosed tissues, it is not improbable that the carbon, and other elements which it contains, are derived from the same source.

There cannot be the smallest doubt, that in the carnivora, the constituents of the urine and the bile are derived from the transformation of compounds of proteine; for, except fat, they consume no food but such as contains proteine, or has been formed from that substance. Their food is identical with their blood; and it is a matter of indifference which of the two we select as the starting-point of the chemical developement of the vital metamorphoses.

There can be no greater contradiction, with regard to the nutritive process, than to suppose that the nitrogen of the food can pass into the urine as urea, without having previously become part of an organized tissue; for albumen, the only constituent of blood, which, from its amount, ought to be taken into consideration, suffers not the slightest change in passing through the liver or kidneys; we find it in every part of the body with the same appearance and the same properties. These organs cannot be adapted for the alteration or decomposition of the substance from which all the other organs of the body are to be formed.

41. From the characters of chyle and lymph, it appears with certainty that the soluble parts of the food or of the chyme acquire the form of albumen. Hard-boiled white of egg, boiled or coagulated fibrine, which have again become soluble in the stomach, but have lost their coagulability by the action of air or heat, recover these properties by degrees. In the chyle, the acid re-action of the chyme has already passed into the weak alkaline re-action of the blood; and the chyle, when, after passing through the mesenteric glands, it has reached the thoracic duct, contains albumen coagulable by heat; and, when left to itself, deposits fibrine. All the compounds of proteine, absorbed during the passage of the chyme through the intestinal canal, take the form of albumen, which, as the results of incubation in the fowl's egg testify, contains the fundamental elements of all organized tissues, with the exception of iron, which is obtained from other sources.

Practical medicine has long ago answered the question, what becomes in man of the compounds of proteine taken in excess, what change is undergone by the superabundant nitrogenised food? The blood-vessels are distended with excess of blood, the other vessels with excess of their fluids, and if the too great supply of food be kept up, and the blood, or other fluids adapted for forming blood, be not applied to their natural purposes, if the soluble matters be not taken up by the proper organs, various gases are disengaged, as in processes of putrefaction, the excrements assume an altered quality in colour, smell, &c. Should the fluids in the absorbent and lymphatic vessels undergo a similar decomposition, this is immediately visible in the blood, and the nutritive process then assumes new forms.

42. No one of all these appearances should occur, if the liver and kidneys were capable of effecting the resolution of the superabundant compounds of proteine into urea, uric acid, and bile. All the observations which have been made in reference to the influence of nitrogenised food on the composition of the urine have failed entirely to demonstrate the existence of any direct influence of the kind; for the phenomena are susceptible of another and a far more simple interpretation, if, along with the food, we consider the mode of life and habits of the individuals who have been the subjects of investigation. Gravel and calculus occur in persons who use very little animal food. Concretions of uric acid have never yet been observed in carnivorous mammalia, living in the wild state,* and among nations which live

---

* The occurrence of urate of ammonia in a concretion found in a dog, which was examined by Lassaigne, is to be doubted, unless Lassaigne extracted it himself from the bladder of the animal.

entirely on flesh, deposits of uric acid concretions in the limbs or in the bladder are utterly unknown.

43. That which must be viewed as an undeniable truth in regard to the origin of the bile, or, more accurately speaking, of choleic acid in the carnivora, cannot hold in regard to all the constituents of the bile secreted by the liver in the herbivora, for with the enormous quantity of bile produced, for example, by the liver of an ox, it is absolutely impossible to suppose that all its carbon is derived from the metamorphosed tissues.

Assuming the 59 oz. of dry bile (from 37 lbs. of fresh bile secreted by an ox) to contain the same percentage of nitrogen as choleic acid (3·86 per cent.), this would amount to nearly $2\frac{1}{4}$ oz. of nitrogen ; and if this nitrogen proceed from metamorphosed tissues, then, if all the carbon of these tissues passed into the bile, it would yield, at the utmost, a quantity of bile corresponding to 7·15 oz. of carbon. This is, however, far below the quantity which, according to observation, is secreted in this class of animals.

44. Other substances, besides compounds of proteine, must inevitably take part in the formation of bile in ‑ the organism of the herbivora ; and these substances can only be the non-nitrogenised constituents of their food.

45. The sugar of bile of Gmelin (picromel or biline of Berzelius), which Berzelius considers as the chief constituent of bile, while Demarçay assigns that place essentially to choleic acid, burns,

when heated in the air, like resin, yields ammoniacal products, and when treated with acids, yields taurine and the products of the decomposition of *choleic acid;* when acted on by alkalies, it yields ammonia and *cholic acid.* At all events, the sugar of bile contains nitrogen, and much less oxygen than starch or sugar, but more oxygen than the oily acids. When, in the metamorphosis of sugar of bile or choleic acid by alkalies, we cause the separation of the nitrogen, we obtain a crystallized acid, very similar to the oily acids (cholic acid), and capable of forming with bases salts, which have the general characters of soaps. Nay, we may even consider the chief constituents of the bile, sugar of bile and choleic acid, as compounds of oily acids with organic oxides, like the fat oils, and only differing from these in containing no oxide of glycerule. Choleic acid, for example, may be viewed as a compound of choloidic acid with allantoine and water :

Choloidic acid.   Allantoine.   Water.   Choleic acid.

$$C_{72}H_{56}O_{12} + C_4N_2H_3O_3 + H_7O_7 = C_{76}N_2H_{66}O_{22}$$

Or as a compound of cholic acid, urea, and water :

Cholic acid.   Urea.   Water.   Choleic acid.

$$C_{74}H_{60}O_{18} + C_2N_2H_4O_2 + H_2O_2 = C_{76}N_2H_{66}O_{22}$$

46. If, in point of fact, as can hardly be doubted, the elements of such substances as starch, sugar, &c., take part in the production of bile in the organism of the herbivora, there is nothing opposed to such a view in the composition of the chief

constituents of bile, as far as our knowledge at
present extends.

If starch be the chief agent in this process, it
can happen in no other way but this—that, as when
it passes into fat, a certain quantity of oxygen is
separated from the elements of the starch, which,
for the same amount of carbon (for 72 atoms), contains five times as much oxygen as choloidic acid.

Without the separation of oxygen from the elements of starch, it is impossible to conceive its
conversion into bile; and this separation being admitted, its conversion into a compound intermediate in composition between starch and fat offers
no difficulty.

47. Not to render these considerations a mere
idle play with formulæ, and not to lose sight of our
chief object, we observe, therefore, that the consideration of the quantitative proportion of the bile
secreted in the herbivora leads to the following
conclusions :—

The chief constituents of the bile of the herbivora contain nitrogen, and this nitrogen is derived
from compounds of proteine.

The bile of this class of animals contains more
carbon than corresponds to the quantity of nitrogenised food taken, or to the portion of tissue that
has undergone metamorphosis in the vital process.

A part of this carbon must, therefore, be derived
from the non-nitrogenised parts of the food (starch,
sugar, &c.); and in order to be converted into a

nitrogenised constituent of bile, a part of the elements of these bodies must necessarily have combined with a nitrogenised compound derived from a compound of proteine.

In reference to this conclusion, it is quite indifferent whether that compound of proteine be derived from the food or from the tissues of the body.

48. It has very lately been stated by A. Ure, that benzoic acid, when administered internally, appears in the urine in the form of hippuric acid.

Should this observation be confirmed,* it will acquire great physiological significance, since it would plainly prove that the act of transformation of the tissues in the animal body, under the influence of certain matters taken in the food, assumes a new form with respect to the products which are its result; for hippuric acid contains the elements of lactate of urea, with the addition of those of benzoic acid:

$$
\left.
\begin{array}{ll}
1 \text{ at. urea} \ldots\ldots\ldots & C_2\,N_2H_4\,O_2 \\
1 \text{ at. lactic acid} \ldots & C_6 \quad H_4\,O_4 \\
2 \text{ at. benzoic acid} & C_{28} \quad H_{10}O_6
\end{array}
\right\}
=
\left\{
\begin{array}{l}
2 \text{ at. crystallized hippuric acid} \\
= 2\,(C_{18}NH_9O_6)
\end{array}
\right.
$$

$$C_{36}N_2H_{18}O_{12}$$

49. If we consider the act of transformation of the tissues in the herbivora as we have done in the

* The analysis of the crystals deposited from the urine on the addition of muriatic acid has not been performed. Besides, the statement of A. Ure, that hippuric acid, dissolved in nitric acid, is reddened by ammonia, is erroneous, and shews that the crystals he obtained must have contained uric acid.

carnivora, then the blood of the former must yield, as the last products of the metamorphosis, from all the organs taken together, choleic acid, uric acid, and ammonia (see p. 136); and if we ascribe to the uric acid an action similar to that of the benzoic acid in Ure's observation—such, namely, that the further transformation, owing to the presence of this acid, assumes another form, the elements of the uric acid being incorporated in the final products— it will appear, for example, that 2 at. of proteine, with the addition of the elements of 3 at. of uric acid and 2 at. of oxygen, might give rise to the production of hippuric acid and urea.

$$2 \text{ at. proteine, } 2 \ (C_{48}N_6H_{36}O_{14}) = C_{96}\,N_{12}H_{72}O_{28}$$
$$3 \text{ at. uric acid, } 3 \ (C_{10}N_4H_4\,O_6\,) = C_{30}\,N_{12}H_{12}O_{18}$$
$$2 \text{ at. oxygen} \qquad\qquad = \qquad O_2$$

$$\text{The sum is } \ldots\ldots\ldots\ldots = C_{126}N_{24}H_{84}O_{48} =$$
$$= \begin{cases} 6 \text{ at. hippuric acid, } 6 \ (C_{18}N\,H_8O_5) = C_{108}N_6\,H_{48}O_{30} \\ 9 \text{ at. urea } \ldots\ldots\ldots 9 \ (C_2\,N_2H_4O_2) = C_{18}\,N_{18}H_{36}O_{33} \end{cases}$$

$$\text{The sum is } \ldots\ldots\ldots \ldots\ldots = C_{126}N_{24}H_{84}O_{48}$$

50. Finally, if we bear in mind, that, in the herbivora, the non-nitrogenised constituents of their food (starch, &c.) must, as we have shewn, play an essential part in the formation of the bile; that to their elements must of necessity be added those of a nitrogenised compound, in order to produce the nitrogenised constituents of the bile, the most striking result of the combinations thus suggested is this, that the elements of starch added to those of

hippuric acid are equal to the elements of choleic acid, *plus* a certain quantity of carbonic acid:

$$\text{2 at. hippuric acid, } 2\,(C_{18}NH_8\,O_5\,) = C_{36}N_2H_{16}O_{10}$$
$$\text{5 at. starch} \ldots\ldots 5\,(C_{12}\;H_{10}O_{10}) = C_{60}\quad H_{50}O_{50}$$
$$\text{2 at. oxygen} \ldots\ldots = O_2$$

The sum is $\ldots\ldots = C_{96}N_2H_{66}O_{62}$

$$=\begin{cases}\text{2 at. choleic acid}\quad 2\,(C_{38}NH_{33}O_{11}) = C_{76}N_2H_{66}O_{22}\\\text{20 at. carbonic acid 20 }(C\quad O_2\,) = C_{20}\quad O_{40}\end{cases}$$

The sum is $\ldots\ldots = C_{96}N_2H_{66}O_{62}$

51. Now, since hippuric acid may be derived, along with urea, from the compounds of proteine, when to the elements of the latter are added those of uric acid (see p. 151); since, further, uric acid, choleic acid, and ammonia contain the elements of proteine in a proportion almost identical with that of proteine itself (see p. 136); it is obvious that, if from 5 at. of proteine, with the addition of oxygen and of the elements of water, there be removed the elements of choleic acid and ammonia, the remainder will represent the elements of hippuric acid and of urea; and that if, when this separation occurs, and during the further transformation, the elements of starch be present and enter into the new products, we shall obtain an additional quantity of choleic acid, as well as a certain amount of carbonic acid gas.

That is to say—that if the elements of proteine and starch, oxygen and water being also present, undergo transformation together and mutually affect each other, we obtain, as the products of this metamorphosis, urea, choleic acid, ammonia, and carbonic acid, and besides these, no other product whatever.

The elements of

$$\left.\begin{array}{l}\text{5 at. proteine}\\\text{15 at. starch}\\\text{12 at. water}\\\text{5 at. oxygen}\end{array}\right\} = \left\{\begin{array}{l}\text{9 at. choleic acid}\\\text{9 at. urea}\\\text{3 at. ammonia}\\\text{60 at. carbonic acid}\end{array}\right.$$

In detail—

5 at. proteine,	5 $(C_{48}N_6H_{36}O_{14})$	=	$C_{240}N_{30}H_{180}O_{70}$
15 at. starch,	15 $(C_{12} \ H_{10}O_{10})$	=	$C_{180} \ H_{150}O_{150}$
12 at. water,	12 (HO)	=	$H_{12} O_{12}$
5 at. oxygen		=	O_5

The sum is .... = $C_{420}N_{30}H_{342}O_{237}$

and—

9 at. choleic acid,	9 $(C_{38}N H_{33}O_{11})$	=	$C_{342}N_9 H_{297}O_{99}$
9 at. urea, ...	9 $(C_2 N_2H_4 O_2)$	=	$C_{18} N_{18}H_{36} O_{18}$
3 at. ammonia, ...	3 (N H_3)	=	$N_3 H_9$
60 at. carbonic acid,	60 $(C \quad O_2)$	=	$C_{60} \quad O_{120}$

The sum is .... = $C_{420}N_{30}H_{342}O_{237}$

The transformation of the compounds of proteine present in the body is effected by means of the oxygen conveyed by the arterial blood, and if the elements of starch, rendered soluble in the stomach, and thus carried to every part, enter into the newly formed compounds, we have the chief constituents of the animal secretions and excretions; carbonic acid, the excretion of the lungs, urea and carbonate of ammonia, excreted by the kidneys, and choleic acid, secreted by the liver.

Nothing, therefore, in the chemical composition of those matters which may be supposed to take a share in these metamorphoses, is opposed to the supposition that a part of the carbon of the non-azotised food enters into the composition of the bile.

52. Fat, in the animal body, disappears when the supply of oxygen is abundant. When that supply is deficient, choleic acid may be converted into hippuric acid, lithofellic acid, (37) and water. Lithofellic acid is known to be the chief constituent of the bezoar stones, which occur in certain herbivorous animals :

$$
\left.\begin{array}{ll}
\text{2 at. choleic acid } C_{76}N_2H_{66}O_{22} \\
\text{10 at. oxygen } \ldots \qquad\qquad O_{10}
\end{array}\right\} = \left\{\begin{array}{ll}
\text{2 at. hip. acid} & C_{36}N_2H_{16}O_{10} \\
\text{1 at. lith. acid} & C_{40} \quad H_{36}O_8 \\
\text{14 at. water } \ldots & H_{14}O_{14}
\end{array}\right.
$$

$$
\overline{C_{76}N_2H_{66}O_{32}} \qquad\qquad\qquad \overline{C_{76}N_2H_{66}O_{32}}
$$

53. For the production of bile in the animal body a certain quantity of soda is, in all circumstances, necessary ; without the presence of a compound of sodium no bile can be formed. In the absence of soda, the metamorphosis of the tissues composed of proteine can yield only fat and urea. If we suppose fat to be composed according to the empirical formula $C_{11}H_{10}O$, then, by the addition of oxygen and the elements of water to the elements of proteine, we have the elements of fat, urea, and carbonic acid.

Proteine.   Water.   Oxygen.

$$
2\,(C_{48}N_6H_{36}O_{14}) + 12\ HO + 14\ O = C_{96}N_{12}H_{84}O_{54} =
$$

$$
= \left\{\begin{array}{ll}
\text{6 at. urea} \ldots\ldots\ldots = C_{12}N_{12}H_{24}O_{12} \\
\text{Fat} \ldots\ldots\ldots\ldots = C_{66} \quad H_{60}O_6 \\
\text{18 at. carbonic acid} = C_{18} \qquad O_{36}
\end{array}\right.
$$

$$
\overline{C_{96}N_{12}H_{84}O_{54}}
$$

The composition of all fats lies between the empirical formulæ $C_{11}H_{10}O$ and $C_{12}H_{10}O$. If we adopt the latter, then the elements of 2 at. proteine, with the addition of 2 at. oxygen and 12 at. water, will yield 6 at. urea, fat ($C_{72}H_{60}O_6$), and 12 at. carbonic acid.

It is worthy of observation, in reference to the production of fat, that the absence of common salt (a compound of sodium which furnishes soda to the animal organism) is favourable to the formation of fat ; that the fattening of an animal is rendered impossible, when we add to its food an excess of salt, although short of the quantity required to produce a purgative effect.

54. As a kind of general view of the metamorphoses of the nitrogenised animal secretions, attention may here be very properly directed to the fact, that the nitrogenised products of the transformation of the bile are identical in ultimate composition with the constituents of the urine, if to the latter be added a certain proportion of the elements of water:

1 at. uric acid	$C_{10}N_4H_4O_6$	$\Big\rbrace =$	$\Big\lbrace$ 3 at. taurine	$C_{12}N_3H_{21}O_{30}$
1 at. urea ...	$C_2N_2H_4O_2$		3 at. ammonia	N_3H_9
22 at. water ...	$H_{22}O_{22}$			
	$C_{12}N_6H_{30}O_{30}$			$C_{12}N_6H_{30}O_{30}$
1 at. allantoine	$C_4N_2H_3O_3$	$\Big\rbrace =$	$\Big\lbrace$ 1 at. taurine	$C_4NH_7O_{10}$
7 at. water ...	H_7O_7		1 at. ammonia	NH_3
	$C_4N_2H_{10}O_{10}$			$C_4N_2H_{10}O_{10}$

55. In reference to the metamorphoses of uric

acid and of the products of the transformation of the bile, it is not less significant, and worthy of remark, that the addition of oxygen and the elements of water to the elements of uric acid may yield either taurine and urea, or taurine, carbonic acid, and ammonia.

$$
\left.\begin{array}{ll}
\text{1 at. uric acid} & C_{10}N_4H_4O_6 \\
\text{14 at. water......} & H_{14}O_{14} \\
\text{2 at. oxygen ...} & O_2
\end{array}\right\} = \left\{\begin{array}{ll}
\text{2 at. taurine} & C_8\,N_2H_{14}O_{20} \\
\text{1 at. urea ...} & C_2\,N_2H\,\,O_2
\end{array}\right.
$$

$$
\left.\begin{array}{ll}
& \overline{C_{10}N_4H_{18}O_{22}} \\
\\
\text{Add 2 at. water} & H_2O_2
\end{array}\right\} = \left\{\begin{array}{ll}
\text{2 at. taurine ...} & C_8\,N_2H_{14}O_{20} \\
\text{2 at. carbon. acid} & C_2\qquad O_4 \\
\text{2 at. ammonia} & N_2H_6
\end{array}\right.
$$

$$ C_{10}N_4H_{20}O_{24} \qquad\qquad\qquad C_{10}N_4H_{20}O_{24} $$

56. Alloxan, *plus* a certain amount of water, is identical in the proportion of elements with taurine; and finally, taurine contains the elements of super-oxalate of ammonia.

$$
\left.\begin{array}{ll}
\text{1 at. alloxan*} & C_8N_2H_4\,O_{10} \\
\text{10 at. water} & H_{10}O_{10}
\end{array}\right\} = \begin{array}{c}\text{Taurine.} \\ 2\,(C_4NH_7O_{10})\end{array}
$$

$$
\text{1 at. taurine } C_4NH_7O_{10} = \left\{\begin{array}{ll}
\text{2 at. oxalic acid} & C_4\qquad O_6 \\
\text{1 at. ammonia} & NH_3 \\
\text{4 at. water ...} & H_4O_4
\end{array}\right.
$$

$$ C_4NH_7O_{10} $$

* It would be most interesting to investigate the action of alloxan on the human body. Two or three drachms, in crystals, had no injurious action on rabbits to which it was given. In man, a large dose appeared to act only on the kidneys. In certain diseases of the liver, alloxan would very probably be found a most powerful remedy.—J. L.

57. The comparison of the amount of carbon in the bile secreted by an herbivorous animal, with the quantity of carbon of its tissues, or of the nitrogenised constituents of its food, which in consequence of the constant transformations may pass into bile, indicates, as we have just seen, a great difference.

The carbon of the bile secreted amounts, at least, to more than five times the quantity of that which could reach the liver in consequence of the change of matter in the body, either from the metamorphosed tissues or from the nitrogenised constituents of the food; and we may regard as well founded the supposition that the non-azotised constituents of the food take a decided share in the production of bile in the herbivora; for neither experience nor observation contradicts this opinion.

58. We have given, in the foregoing paragraphs, the analytical proof, that the nitrogenised products of the transformation of bile, namely, taurine and ammonia, may be formed from all the constituents of the urine, with the exception of urea—that is, from hippuric acid, uric acid, and allantoine; and when we bear in mind that, by the mere separation of oxygen and the elements of water, choloidic acid may be formed from starch;—

From 6 at. starch $= 6 \ (C_{12}H_{10}O_{10}) = C_{72}H_{60}O_{60}$

$$\left.\begin{array}{l}\text{Subtract 44 at. oxygen}\\ \text{4 at. water}\end{array}\right\} = \qquad H_4 O_{48}$$

Remains choloidic acid .... $= C_{72}H_{56}O_{12}$;—

that, finally, choloidic acid, ammonia, and taurine,

if added together, contain the elements of choleic acid;—

$$\begin{array}{lll}
\text{1 at. choloidic acid} & = & C_{72} \quad H_{56} O_{12} \\
\text{1 at. taurine} \ldots\ldots & = & C_4 \, N \, H_7 \, O_{10} \\
\text{1 at. ammonia} \ldots & = & N \, H_3 \\
\hline
\text{1 at. choleic acid} & = & C_{76} N_2 H_{66} O_{22} \,;\!— \\
\end{array}$$

if all this be considered, every doubt as to the possibility of these changes is removed.

59. Chemical analysis and the study of the living animal body mutually support each other; and both lead to the conclusion that a certain portion of the carbon of the non-azotised constituents of food (of starch, &c., the elements of respiration) is secreted by the liver in the form of bile; and further, that the nitrogenised products of the transformation of tissues in the herbivora do not, as in the carnivora, reach the kidneys immediately or directly, but that, before their expulsion from the body in the form of urine, they take a share in certain other processes, especially in the formation of the bile.

They are conveyed to the liver with the non-azotised constituents of the food; they are returned to the circulation in the form of bile, and are not expelled by the kidneys till they have thus served for the production of the most important of the substances employed in respiration.

60. When the urine is left to itself, the urea which it contains is converted into carbonate of ammonia; the elements of urea are in such propor-

tion, that by the addition of the elements of water, all its carbon is converted into carbonic acid, and all its nitrogen into ammonia.

$$
\begin{array}{ll}
\text{1 at. urea} \quad C_2N_2H_4O_2 \\
\text{2 at. water} \quad H_2O_2
\end{array}
\Bigg\} =
\begin{cases}
\text{2 at. carbonic acid } C_2 O_4 \\
\text{2 at. ammonia} \ldots N_2H_6
\end{cases}
$$

$$\overline{C_2N_2H_6O_4} \qquad\qquad\qquad \overline{C_2N_2H_6O_4}$$

61. Were we able directly to produce taurine and ammonia out of uric acid or allantoine, this might perhaps be considered as an additional proof of the share which has been ascribed to these compounds in the production of bile; it cannot, however, be viewed as any objection to the views above developed on the subject, that, with the means we possess, we have not yet succeeded in effecting these transformations out of the body. Such an objection loses all its force, when we consider that we cannot admit, as proved, the pre-existence of taurine and ammonia in the bile; nay, that it is not even probable that these compounds, which are only known to us as products of the decomposition of the bile, exist ready formed, as ingredients of that fluid.

By the action of muriatic acid on bile, we, in a manner, force its elements to unite in such forms as are no longer capable of change under the influence of the same re-agent; and when, instead of the acid, we use potash, we obtain the same elements, although arranged in another, and quite a different manner. If taurine were present, ready formed, in bile, we should obtain the same products by the action of acids and of alkalies. This, however, is contrary to experience.

Thus, even if we could convert allantoine, or uric acid and urea, into taurine and ammonia, out of the body, we should acquire no additional insight into the true theory of the formation of bile, just because the pre-existence of ammonia and taurine in the bile must be doubted, and because we have no reason to believe that urea or allantoine, as such, are employed by the organism in the production of bile. We can prove that their elements serve this purpose, but we are utterly ignorant how these elements enter into these combinations, or what is the chemical character of the nitrogenised compound which unites with the elements of starch to form bile, or rather choleic acid.

62. Choleic acid may be formed from the elements of starch with those of uric acid and urea, or of allantoine, or of uric acid, or of alloxan, or of oxalic acid and ammonia, or of hippuric acid. The possibility of its being produced from so great a variety of nitrogenised compounds is sufficient to shew that all the nitrogenised products of the metamorphosis of the tissues may be employed in the formation of bile, while we cannot tell in what precise way they are so employed.

By the action of caustic alkalies allantoine may be resolved into oxalic acid and ammonia; the same products are obtained when oxamide is acted

on by the same re-agents. Yet we cannot, from
the similarity of the products, conclude that these
two compounds have a similar constitution. In like
manner the nature of the products formed by the
action of acids on choleic acid does not entitle us
to draw any conclusion as to the form in which its
elements are united together.

63. If the problem to be solved by organic chemistry be this, namely, to explain the changes which
the food undergoes in the animal body ; then it is
the business of this science to ascertain what elements must be added, what elements must be separated, in order to effect, or, in general, to render possible, the conversion of a given compound
into a second or a third ; but we cannot expect
from it the synthetic proof of the accuracy of the
views entertained, because every thing in the organism goes on under the influence of the vital force,
an immaterial agency, which the chemist cannot
employ at will.

The study of the phenomena which accompany
the metamorphoses of the food in the organism, the
discovery of the share which the atmosphere or the elements of water take in these changes, lead at once
to the conditions which must be united in order to
the production of a secretion or of an organized part.

64. The presence of free muriatic acid in the
stomach, and that of soda in the blood, prove beyond
all doubt the necessity of common salt for the organic processes ; but the quantities of soda required

M

by animals of different classes, to support the vital
processes, are singularly unequal.

If we suppose, that a given amount of blood,
considered as a compound of soda, passes, in the
body of a carnivorous animal, in consequence of the
change of matter, into a new compound of soda,
namely, the bile, we must assume, that in the normal condition of health, the proportion of soda in
the blood is amply sufficient to form bile with the
products of transformation. The soda which has
been used in the vital processes, and any excess of
soda, must be expelled in the form of a salt, after
being separated from the blood by the kidneys.

Now, if it be true, that, in the body of an herbivorous animal, a much larger quantity of bile is produced than corresponds to the amount of blood
formed or transformed in the vital processes; if the
greater part of the bile, in this case, proceeds from
the non-azotised constituents of the food, then the
soda of the blood which has been formed into organised tissue (assimilated or metamorphosed) cannot possibly suffice for the supply of the daily secretion of bile. The soda, therefore, of the bile of the
herbivora must be supplied directly in the food;
their organism must possess the power of applying
directly to the formation of bile all the compounds
of soda present in the food, and decomposable by
the organic process. All the soda of the animal
-body obviously proceeds from the food; but the
food of the carnivora contains, at most, only the

amount of soda necessary to the formation of blood ; and in most cases, among animals of this class, we may assume that only as much soda as corresponds to the proportion employed to form the blood is expelled in the urine.

When the carnivora obtain in their food as much soda as suffices for the production of their blood, an equal amount is excreted in the urine ; when the food contains less, a part of that which would otherwise be excreted is retained by the organism.

All these statements are most unequivocally confirmed by the composition of the urine in these different classes of animals.

65. As the ultimate product of the changes of all compounds of soda in the animal body, we find in the urine the soda in the form of a salt, and the nitrogen in that of ammonia or urea.

The soda in the urine of the carnivora is found in combination with sulphuric and phosphoric acids ; and along with the sulphate and phosphate of soda we never fail to find a certain quantity of a salt of ammonia, either muriate or phosphate of ammonia. There can be no more decisive evidence in favour of the opinion, that the soda of their bile or of the metamorphosed constituents of their blood is very far from sufficing to neutralize the acids which are separated, than the presence of ammonia in their urine. This urine, moreover, has an acid re-action.

In contradistinction to this, we find, in the urine of the herbivora, soda in predominating quantity ;

and that not' combined with sulphuric or phosphoric
acids, but with carbonic, benzoic, or hippuric acids.

66. These well-established facts demonstrate
that the herbivora consume a far larger quantity
of soda than is required merely for the supply of
the daily consumption of blood. In their food are
united all the conditions for the production of a
second compound of soda, destined for the support
of the respiratory process ; and it can only be a very
limited knowledge of the vast wisdom displayed in
the arrangements of organized nature which can
look on the presence of so much soda in the food
and in the urine of the herbivora as accidental.

It cannot be accidental, that the life, the developement of a plant is dependant on the presence of the
alkalies which it extracts from the soil. This plant
serves as food to an extensive class of animals, and
in these animals the vital process is again most
closely connected with the presence of these alkalies.
We find the alkalies in the bile, and their presence
in the animal body is the indispensable condition
for the production of the first food of the young
animal; for without an abundant supply of potash,
the production of milk becomes impossible.

67. All observation leads, as appears from the
preceding exposition, to the opinion, that certain
non-azotised constituents of the food of the herbivora (starch, sugar, gum, &c.) acquire the form of
a compound of soda, which, in their bodies, serves
for the same purpose as that which we know cer-

tainly to be served by the bile (the most highly car-
bonized product of the transformation of their tissues)
in the bodies of the carnivora. These substances
are employed to support certain vital actions, and
are finally consumed in the generation of animal
heat, and in furnishing means of resistance to the
action of the atmosphere. In the carnivora, the
rapid transformation of their tissues is a condition of
their existence, because it is only as the result of
the change of matter in the body that those substances can be formed, which are destined to enter
into combination with the oxygen of the air ; and
in this sense we may say that the non-azotised constituents of the food of the herbivora impede the
change of matter, or retard it, and render unnecessary, at all events, so rapid a process as occurs in
the carnivora.

68. The quantity of azotised matter, proportion-
ally so small, which the herbivora require to support their vital functions, is closely connected with
the power possessed by the non-azotised parts of
their food to act as means of supporting the respiratory process ; and this consideration seems to
render it not improbable, that the necessity for
more complex organs of digestion in the herbivora
is rather owing to the difficulty of rendering soluble
and available for the vital processes certain non-
azotised compounds (gum ? amylaceous fibre ?) than
to any thing in the change or transformation of
vegetable fibrine, albumen, and caseine into blood;

since, for this latter purpose, the less complex digestive apparatus of the carnivora is amply sufficient.

69. If, in man, when fed on a mixed diet, starch perform a similar part to that which it plays in the body of the herbivora; if it be assumed that the elements of starch are equally necessary to the formation of the bile in man as in these animals; then it follows that a part of the azotised products of the transformation of the tissues in the human body, before they are expelled through the bladder, returns into the circulation from the liver in the shape of bile, and is separated by the kidneys from the blood, as the ultimate product of the respiratory process.

70. When there is a deficiency of non-azotised matter in the food of man, this form of the production of bile is rendered impossible. In that case, the secretions must possess a different composition; and the appearance of uric acid in the urine, the deposition of uric acid in the joints and in the bladder, as well as the influence which an excess of animal food (which must be considered equivalent to a deficiency of starch, &c.) exercises on the separation of uric acid in certain individuals, may be explained on this principle. If starch, sugar, &c., be deficient, then a part of the azotised compounds formed during the change of matter will either remain in the situation where they have been formed, in which case they will not be sent from

the liver into the circulation, and therefore will not undergo the final changes dependant on the action of oxygen; or they will be separated by the kidneys in some form different from the normal one.

71. In the preceding paragraphs I have endeavoured to prove that the non-azotised constituents of food exercise a most decided influence on the nature and quality of the animal secretions. Whether this occur directly; whether, that is to say, their elements take an immediate share in the act of transformation of tissues; or whether their share in that process be an indirect one, is a question probably capable of being resolved by careful and cautious experiment and observation. It is possible, that the non-azotised constituents of food, after undergoing some change, are carried from the intestinal canal directly to the liver, and that they are converted into bile in this organ, where they meet with the products of the metamorphosed tissues, and subsequently complete their course through the circulation.

This opinion appears more probable, when we reflect that as yet no trace of starch or sugar has been detected in arterial blood, not even in animals which had been fed exclusively with these substances. We cannot ascribe to these substances, since they are wanting in arterial blood, any share in the nutritive process; and the occurrence of sugar in the urine of those affected with diabetes mellitus (which sugar, according to the best obser-

vations, is derived from the food) coupled with its
total absence in the blood of the same patients, obviously proves that starch and sugar are not, as such,
taken into the circulation.

72. The writings of physiologists contain many
proofs of the presence of certain constituents of the
bile in the blood of man in a state of health, although their quantity can hardly be determined.
Indeed, if we suppose $8\frac{1}{4}$ lbs. (58,000 grs.) of blood
to pass through the liver every minute, and if from
this quantity of blood 2 drops of bile (3 grains to
the drop) are secreted, this would amount to $\frac{1}{9600}$th
part of the weight of the blood, a proportion far too
small to be quantitatively ascertained by analysis.

73. The greater part of the bile in the body of
the herbivora, and in that of man fed on mixed
food, appears from the preceding considerations to
be derived from the elements of the non-azotised
food. But its formation is impossible without the
addition of an azotised body, for the bile is a compound of nitrogen. All varieties of bile yet examined yield, when subjected to dry distillation,
ammonia and other nitrogenised products. Taurine
and ammonia may easily be extracted from ox bile ;
and the only reason why we cannot positively prove
that the same products may be obtained from the
bile of other animals is this, that it is not easy to
procure, in the case of many of these animals, a
sufficient quantity of bile for the experiment.

Now, whether the nitrogenised compound which

unites with the elements of starch to form bile be
derived from the food or from the substance of the
metamorphosed tissues, the conclusion that its presence is an essential condition for the secretion of
bile cannot be considered doubtful.

Since the herbivora obtain in their food only such
nitrogenised compounds as are identical in composition with the constituents of their blood, it is at all
events clear, that the nitrogenised compound which
enters into the composition of bile is derived from
a compound of proteine. It is either formed in
consequence of a change which the compounds of
proteine in the food have undergone, or it is produced from the blood or from the substance of the
tissues by the act of their metamorphosis.

74. If the conclusion be accurate, that nitrogenised compounds, whether derived from the blood or
from the food, take a decided share in the formation
of the secretions, and particularly of the bile, then
it is plain that the organism must possess the power
of causing foreign matters, which are neither parts
nor constituents of the organs in which vital activity
resides, to serve for certain vital processes. All nitrogenised substances capable of being rendered soluble,
without exception, when introduced into the organs
of circulation or of digestion, must, if their composition be adapted for such purposes, be employed by
the organism in the same manner as the nitrogenised products which are formed in the act of metamorphosis of tissues.

We are acquainted with a multitude of substances, which exercise a most marked influence on the act of transformation as well as on the nutritive process, while their elements take no share in the resulting changes. These are uniformly substances the particles of which are in a certain state of motion or decomposition, which state is communicated to all such parts of the organism as are capable of undergoing a similar transformation.

75. Medicinal and poisonous substances form a second and most extensive class of compounds, the elements of which are capable of taking a direct or an indirect share in the processes of secretion and of transformation. These may be subdivided into three great orders; the first (which includes the metallic poisons) consists of substances which enter into chemical combination with certain parts or constituents of the body, while the vital force is insufficient to destroy the compounds thus formed. The second division, consisting of the essential oils, camphor, empyreumatic substances, and antiseptics, &c., possesses the property of impeding or retarding those kinds of transformation to which certain very complex organic molecules are liable; transformations which, when they take place out of the body, are usually designated by the names of fermentation and putrefaction.

The third division of medicinal substances is composed of bodies, the elements of which take a direct share in the changes going on in the animal body. When introduced into the system, they augment the energy of the vital activity of one or more organs; they excite morbid phenomena in the healthy body. All of them produce a marked effect in a comparatively small dose, and many are poisonous when administered in larger quantity. None of the substances in this class can be said to take a decided share in the nutritive process, or to be employed by the organism in the production of blood; partly, because their composition is different from that of blood, and, partly, because the proportion in which they must be given, to exert their influence, is as nothing, compared with the mass of the blood.

These substances, when taken into the circulation, alter, as is commonly said, the quality of the blood, and in order that they may pass from the stomach into the circulation with their entire efficacy, we must assume that their composition is not affected by the organic influence of the stomach. If insoluble when given, they are rendered soluble in that organ, but they are not decomposed; otherwise, they would be incapable of exerting any influence on the blood.

76. The blood, in its normal state, possesses two qualities closely related to each other, although we may conceive one of them to be quite independent of the other.

The blood contains, in the form of the globules, the carriers, as it were, of the oxygen which serves for the production of certain tissues, as well as for

the generation of animal heat. The globules of
the blood, by means of the property they possess
of giving off the oxygen they have taken up in
the lungs, without losing their peculiar character,
determine generally the change of matter in the
body.

The second quality of the blood, namely, the
property which it possesses of becoming part of an
organised tissue, and its consequent adaptation to
promote the formation and the growth of organs, as
well as to the reproduction or supply of waste in
the tissues, is owing, chiefly, to the presence of dissolved fibrine and albumen. These two chief constituents, which serve for nutrition and reproduction of matter, in passing through the lungs are
saturated with oxygen, or, at all events, absorb so
much from the atmosphere as entirely to lose the
power of extracting oxygen from the other substances present in the blood.

77. We know for certain that the globules of
the venous blood, when they come in contact with
air in the lungs, change their colour, and that this
change of colour is accompanied by an absorption of
oxygen; and that all those constituents of the blood,
which possess in any degree the power of combining
with oxygen, absorb it in the lungs, and become saturated with it. Although in contact with these
other compounds, the globules, when arterialised,
retain their florid, red colour in the most minute
ramifications of the arteries; and we observe them

to change their colour, and to assume the dark red
colour which characterizes venous blood, only during
their passage through the capillaries. From these
facts we must conclude that the constituents of
arterial blood are altogether destitute of the power
to deprive the arterialised globules of the oxygen
which they have absorbed from the air; and we
can draw no other conclusion from the change of
colour which occurs in the capillaries, than that the
arterialised globules, during their passage through
the capillaries, return to the condition which cha-
racterized them in venous blood; that, consequently,
they give up the oxygen absorbed in the lungs, and
thus acquire the power of combining with that
element afresh.

78. We find, therefore, in arterial blood, albumen, which, like all the other constituents of that
fluid, has become saturated with oxygen in its passage through the lungs, and oxygen gas, which is
conveyed to every particle in the body in chemical
combination with the globules of the blood. As
far as our observations extend (in the developement
of the chick during incubation), all the conditions
seem to be here united which are necessary to the
formation of every kind of tissue; while that portion of oxygen which is not consumed in the growth
or reproduction of organs combines with the substance of the living parts, and produces, by its
union with their elements, the act of transformation which we have called the change of matter.

79. It is obvious, that all compounds, of whatever kind, which are present in the capillaries, whether separated there, or introduced by endosmosis or imbibition, if not altogether incapable of uniting with oxygen, must, when in contact with the arterialised globules, the carriers of oxygen, be affected exactly in the same way as the solids forming part of living organs. These compounds, or their elements, will enter into combination with oxygen, and in this case there will either be no change of matter, or that change will exhibit itself in another form, yielding products of a different kind.

80. The conception, then, of a change in the two qualities of the blood above alluded to, by means of a foreign body contained in the blood or introduced into the circulation (a medicinal agent), presupposes two kinds of operation.

Assuming that the remedy cannot enter into any such chemical union with the constituents of the blood as puts an end to the vital activity; assuming, further, that it is not in a condition of transformation capable of being communicated to the constituents of the blood or of the organs, and of continuing in them; assuming, lastly, that it is incapable, by its contact with the living parts, of putting a stop to the change of matter, the transformation of their elements; then, in order to discover the modus operandi of this class of medicinal agents, nothing is left but to conclude that their elements take a share in the formation of certain constituents

of the living body, or in the production of certain
secretions.

81. The vital process of secretion, in so far as it
is related to the chemical forces, has been subjected
to examination in the preceding pages. In the carnivora we have reason to believe, that, without the
addition of any foreign matter in the food, the bile
and the constituents of the urine are formed in
those parts where the change of matter takes place.
In other classes of animals, on the other hand, we
may suppose that in the organ of secretion itself,
the secreted fluid is produced from certain matters
conveyed to it; in the herbivora, for example, the
bile is formed from the elements of starch along
with those of a nitrogenised product of the metamorphosis of the tissues. But this supposition by
no means excludes the opinion, that in the carnivora the products of the metamorphosed tissues are
resolved into bile, uric acid, or urea, only after reaching the secreting organ; nor the opinion that the
elements of the non-azotised food, conveyed directly
by the circulation to every part of the body, where
change of matter is going on, may there unite with
the elements of the metamorphosed tissues, to form
the constituents of the bile and of the urine.

82. If we now assume, that certain medicinal
agents may become constituents of secretions, this
can only occur in two ways.   Either they enter the
circulation, and take a direct share in the change of
matter, in so far as their elements enter into the

composition of the new products; or they are conveyed to the organs of secretion, where they exert
an influence on the formation or on the quality of a
secretion by the addition of their elements.

In either case, they must lose in the organism
their chemical character; and we know with sufficient certainty, that this class of medicinal bodies
disappears in the body without leaving a trace.   In
fact, if we ascribe to them any effect, they cannot
lose their peculiar character by the action of the
stomach; their disappearance, therefore, presupposes
that they have been applied to certain purposes,
which cannot be imagined to occur without a change
in their composition.

83. Now, however limited may be our knowledge
of the composition of the different secretions, with
the exception of the bile, this much is certain, that
all the secretions contain nitrogen chemically combined.   They pass into fetid putrefaction, and yield
either in this change, or in the dry distillation, ammoniacal products.   Even the saliva, when acted on
by caustic potash, disengages ammonia freely.

84. Medicinal or remedial agents may be divided
into two classes, the nitrogenised and the non-nitrogenised.   The nitrogenised vegetable principles,
whose composition differs from that of the proper
nitrogenised elements of nutrition, also produced by
a vegetable organism, are distinguished, beyond all
others, for their powerful action on the animal economy.

The effects of these substances are singularly varied; from the mildest form of the action of aloes, to the most terrible poison, strychnia, we observe an endless variety of different actions.

With the exception of three, all these substances produce diseased conditions in the healthy organism, and are poisonous in certain doses. Most of them are, chemically speaking, basic or alkaline.

No remedy, devoid of nitrogen, possesses a poisonous action in a similar dose.*

85. The medicinal or poisonous action of the nitrogenised vegetable principles has a fixed relation to their composition; it cannot be supposed to be independent of the nitrogen they contain, but is certainly not in direct proportion to the quantity of nitrogen.

Solanine (38), and picrotoxine (39), which contain least nitrogen, are powerful poisons. Quinine (40) contains more nitrogen than morphia (41). Caffeine (42), and theobromine, the most highly nitrogenised of all vegetable principles, are not poisonous.

86. A nitrogenised body, which exerts, by means of its elements, an influence on the formation or on the quality of a secretion, must, in regard to its

* This consideration or comparative view has led lately to a more accurate investigation of the composition of picrotoxine, the poisonous principle of cocculus indicus; and M. Francis has discovered the existence of nitrogen in it, hitherto overlooked, and has also determined its amount.

chemical character, be capable of taking the same share as the nitrogenised products of the animal body do in the formation of the bile; that is, it must play the same part as a product of the vital process. On the other hand, a non-azotised medicinal agent, in so far as its action affects the secretions, must be capable of performing in the animal body the same part as that which we have ascribed in the formation of the bile, to the non-azotised elements of food.

Thus, if we suppose that the elements of hippuric or uric acids are derived from the substance of the organs in which vitality resides; that, as products of the transformation of these organs, they lose the vital character, without losing the capacity of undergoing changes under the influence of the inspired oxygen, or of the apparatus of secretion; we can hardly doubt that similar nitrogenised compounds, products of the vital process in plants, when introduced into the animal body, may be employed by the organism exactly in the same way as the nitrogenised products of the metamorphosis of the animal tissues themselves. If hippuric and uric acids, or any of their elements, can take a share, for example, in the formation and supply of bile, we must allow the same power to other analogous nitrogenised compounds.

We shall never, certainly, be able to discover how men were led to the use of the hot infusion of the leaves of a certain shrub (tea), or of a decoction

of certain roasted seeds (coffee). Some cause there
must be, which would explain how the practice has
become a necessary of life to whole nations. But
it is surely still more remarkable, that the beneficial
effects of both plants on the health must be ascribed
to one and the same substance, the presence of
which in two vegetables, belonging to different
natural families, and the produce of different quarters of the globe, could hardly have presented itself
to the boldest imagination. Yet recent researches
have shewn, in such a manner as to exclude all
doubt, that caffeine, the peculiar principle of coffee,
and theine, that of tea, are, in all respects, identical.

It is not less worthy of notice, that the American
Indian, living entirely on flesh, discovered for himself, in tobacco smoke, a means of retarding the
change of matter in the tissues of his body, and
thereby of making hunger more endurable; and
that he cannot withstand the action of brandy,
which, acting as an element of respiration, puts
a stop to the change of matter by performing the
function which properly belongs to the products of
the metamorphosed tissues. Tea and coffee were
originally met with among nations whose diet is
chiefly vegetable.

87. Without entering minutely into the medicinal action of caffeine (theine), it will surely appear
a most striking fact, even if we were to deny its
influence on the process of secretion, that this substance, with the addition of oxygen and the elements

of water, can yield taurine, the nitrogenised compound peculiar to bile :

$$1 \text{ at. caffeine or theine} = C_8N_2H_5\,O_2$$
$$9 \text{ at. water} \dots\dots\dots = \quad H_9\,O_9$$
$$9 \text{ at. oxygen} \dots\dots\dots = \quad\quad O_9$$
$$\overline{ C_8N_2H_{14}O_{20}} =$$
$$= 2 \text{ at. taurine} \dots\dots\dots = 2\,(C_4NH_7O_{10})$$

A similar relation exists in the case of the peculiar principle of asparagus and of althæa, asparagine; which also, by the addition of oxygen and the elements of water, yields the elements of taurine :

$$1 \text{ at. asparagine} = C_8N_2H_8\,O_6$$
$$6 \text{ at. water} \dots\dots = \quad H_6\,O_6$$
$$8 \text{ at. oxygen} \dots = \quad\quad O_8$$
$$\overline{ C_8N_2H_{14}O_{20}} =$$
$$= 2 \text{ at. taurine} = 2\,(C_4NH_7O_{10})$$

The addition of the elements of water and of a certain quantity of oxygen to the elements of theobromine, the characteristic principle of the cacao bean (theobroma cacao), yields the elements of taurine and urea, of taurine, carbonic acid, and ammonia, or of taurine and uric acid :

$$
\left.
\begin{array}{l}
1 \text{ at. theobromine } C_{18}N_6H_{10}O_4 \\
22 \text{ at. water } \dots\dots \quad H_{22}O_{22} \\
16 \text{ at. oxygen } \dots \quad\quad O_{16}
\end{array}
\right\}
=
\left\{
\begin{array}{l}
4 \text{ at. taurine } C_{16}N_4H_{28}O_{40} \\
1 \text{ at. urea} \dots C_2\,N_2H_4\,O_2
\end{array}
\right.
$$

$$\overline{C_{18}N_6H_{32}O_{42}} \qquad\qquad \overline{C_{18}N_6H_{32}O_{42}}$$

or—

$$
\left.
\begin{array}{l}
1 \text{ at. theobromine } C_{18}N_6\,H_{10}O_4 \\
24 \text{ at. water } \dots\dots \quad H_{24}O_{24} \\
16 \text{ at. oxygen} \dots\dots \quad\quad O_{16}
\end{array}
\right\}
=
\left\{
\begin{array}{l}
4 \text{ at. taurine } \quad C_{16}N_4\,H_{28}O_{40} \\
2 \text{ at. carb. acid } C_2 \quad\quad O_4 \\
2 \text{ at. ammonia } \quad N_2\,H_6
\end{array}
\right.
$$

$$\overline{C_{18}N_6H_{34}\,O_{44}} \qquad\qquad \overline{C_{18}N_6\,H_{34}O_{44}}$$

or—

1 at. theobromine	$C_{18}N_6H_{10}O_4$			2 at. taurine	$C_8 N_2 H_{14}O_{20}$
8 at. water	$H_8 O_8$	$=$		1 at. uric acid	$C_{10}N_4H_4 O_6$
14 at. oxygen	O_{14}				
	$C_{18}N_6H_{18}O_{26}$				$C_{18}N_6H_{18}O_{26}$

88. To see how the action of caffeine, asparagine, theobromine, &c., may be explained, we must call to mind that the chief constituent of the bile contains only 3·8 per cent. of nitrogen, of which only the half, or 1·9 per cent., belongs to the taurine.

Bile contains, in its natural state, water and solid matter, in the proportion of 90 parts by weight of the former to 10 of the latter. If we suppose these 10 parts by weight of solid matter to be choleic acid, with 3·87 per cent. of nitrogen, then 100 parts of fresh bile will contain 0·171 parts of nitrogen in the shape of taurine. Now this quantity is contained in 0·6 parts of caffeine; or $2\frac{8}{10}$ths grains of caffeine can give to an ounce of bile the nitrogen it contains in the form of taurine. If an infusion of tea contain no more than the $\frac{1}{70}$th of a grain of caffeine, still, if it contribute in point of fact to the formation of bile, the action, even of such a quantity, cannot be looked upon as a nullity. Neither can it be denied that in the case of an excess of non-azotised food and a deficiency of motion, which is required to cause the change of matter in the tissues, and thus to yield the nitrogenised product which enters into the composition of the bile; that

in such a condition, the health may be benefited by
the use of compounds which are capable of supplying the place of the nitrogenised product produced in the healthy state of the body, and essential to the production of an important element of
respiration.   In a chemical sense—and it is this
alone which the preceding remarks are intended to
shew—caffeine or theine, asparagine, and theobromine, are, in virtue of their composition, better
adapted to this purpose than all other nitrogenised vegetable principles.   The action of these substances, in ordinary circumstances, is not obvious,
but it unquestionably exists.

89. With respect to the action of the other nitrogenised vegetable principles, such as quinine, or the
alkaloids of opium, &c., which manifests itself, not in
the processes of secretion, but in phenomena of another kind, physiologists and pathologists entertain
no doubt that it is exerted chiefly on the brain and
nerves.   This action is commonly said to be dyna-
mic—that is, it accelerates, or retards, or alters in
some way the phenomena of motion in animal life.
If we reflect that this action is exerted by substances which are material, tangible and ponder-
able; that they disappear in the organism; that a
double dose acts more powerfully than a single one;
that, after a time, a fresh dose must be given, if we
wish to produce the action a second time; all these
considerations, viewed chemically, permit only one
form of explanation; the supposition, namely, that

these compounds, by means of their elements, take a share in the formation of new, or the transformation of existing brain and nervous matter.

However strange the idea may, at first sight, appear, that the alkaloids of opium or of cinchona bark, the elements of codeine, morphia, quinine, &c., may be converted into constituents of brain and nervous matter, into organs of vital energy, from which the organic motions of the body derive their origin; that these substances form a constituent of that matter, by the removal of which the seat of intellectual life, of sensation, and of consciousness, is annihilated: it is, nevertheless, certain, that all these forms of power and activity are most closely dependant, not only on the existence, but also on a certain quality of the substance of the brain, spinal marrow, and nerves; insomuch, that all the manifestations of the life or vital energy of these modifications of nervous matter, which are recognized as the phenomena of motion, sensation, or feeling, assume another form as soon as their composition is altered. The animal organism has produced the brain and nerves out of compounds furnished to it by vegetables; it is the constituents of the food of the animal, which, in consequence of a series of changes, have assumed the properties and the structure which we find in the brain and nerves.

90. If it must be admitted as an undeniable truth, that the substance of the brain and nerves is

produced from the elements of vegetable albumen,
fibrine and caseine, either alone, or with the aid of
the elements of non-azotised food, or of the fat
formed from the latter, there is nothing absurd in
the opinion, that other constituents of vegetables,
intermediate in composition between the fats and
the compounds of proteine, may be applied in the
organism to the same purpose.

91. According to the researches of Fremy, the
chief constituent of the fat found in the brain is a
compound of soda with a peculiar acid, the *cerebric
acid*, which contains, in 100 parts,

Carbon ...	...	...	...	...	66·7
Hydrogen	...	...	...	...	10·6
Nitrogen	...	...	...	...	2·3
Phosphorus	...	...	...	...	0·9
Oxygen ...	...	...	...	...	19·5

It is easy to see that the composition of cerebric
acid differs entirely, both from that of ordinary fats
and of the compounds of proteine.   Common fats
contain no nitrogen, while the compounds of proteine contain nearly 17 per cent.   Leaving the
phosphorus out of view, the composition of this
acid approaches most nearly to that of choleic acid,
although these two compounds are quite distinct.

92. Brain and nervous matter is, at all events,
formed in a manner similar to that in which bile is
produced ; either by the separation of a highly nitrogenised compound from the elements of blood, or
by the combination of a nitrogenised product of the

vital process with a non-azotised compound (probably, a fatty body). All that has been said in the
preceding pages on the various possible ways by
which the bile might be supposed to be formed, all
the conclusions which we attained in regard to the
co-operation of azotised and non-azotised elements
of food, may be applied with equal justice and equal
probability to the formation and production of the
nervous substance.

We must not forget that, in whatever light we
may view the vital operations, the production of
nervous matter from blood presupposes a change in
the composition and qualities of the constituents of
blood. That such a change occurs is as certain as
that the existence of the nervous matter cannot be
denied. In this sense, we must assume, that from
a compound of proteine may be formed a first, second, third, &c., product, before a certain number of
its elements can become constituents of the nervous
matter ; and it must be considered as quite certain,
that a product of the vital process in a plant, introduced into the blood, will, if its composition be
adapted to this purpose, supply the place of the first,
second, or third product of the alteration of the
compound of proteine. Indeed it cannot be considered merely accidental, that the composition of the
most active remedies, namely, the vegetable alkaloids, cannot be shewn to be related to that of any
constituent of the body, except only the substance
of the nerves and brain. All of these contain a

certain quantity of nitrogen, and, in regard to their composition, they are intermediate between the compounds of proteine and the fats.

93. In contradistinction to their chemical character, we find that the substance of the brain exhibits the characters of an acid. It contains far more oxygen than the organic bases or alkaloids. We observe, that quinine and cinchonine, morphia and codeine, strychnia and brucia, which are, respectively, so nearly alike in composition, if they do not produce absolutely the some effect, yet resemble each other in their action more than those which differ more widely in composition. We find that their energy of action diminishes, as the amount of oxygen they contain increases (as in the case of narcotine), and that, strictly speaking, no one of them can be entirely replaced by another. There cannot be a more decisive proof of the nature of their action than this last fact; it must stand in the closest relation to their composition. If these compounds, in point of fact, are capable of taking a share in the 'formation or in the alteration of the qualities of brain and nervous matter, their action on the healthy as well as the diseased organism admits of a surprisingly simple explanation. If we are not tempted to deny, that the chief constituent of soup may be applied to a purpose corresponding to its composition in the human body, or that the organic constituent of bones may be so employed in the body of the dog, although that substance (gelatine

in both cases) is absolutely incapable of yielding blood; if, therefore, nitrogenised compounds, totally different from the compounds of proteine, may be employed for purposes corresponding to their composition; we may thence conclude that a product of vegetable life, also different from proteine, but similar to a constituent of the animal body, may be employed by the organism in the same way and for the same purpose as the natural product, originally formed by the vital energy of the animal organs, and that, indeed, from a vegetable substance.

The time is not long gone by, when we had not the very slightest conception of the cause of the various effects of opium, and when the action of cinchona bark was shrouded in incomprehensible obscurity. Now that we know that these effects are caused by crystallizable compounds, which differ as much in composition as in their action on the system; now that we know the substances to which the medicinal or poisonous energy must be ascribed, it would argue only want of sense to consider the action of these substances inexplicable; and to do so, as many have done, because they act in very minute doses, is as unreasonable as it would be to judge of the sharpness of a razor by its weight.

94. It would serve no purpose to give these considerations a greater extension at present. However hypothetical they may appear, they only deserve attention in so far as they point out the way which chemistry pursues, and which she ought not

to quit, if she would really be of service to physiology and pathology. The combinations of the chemist relate to the change of matter, forwards and backwards, to the conversion of food into the various tissues and secretions, and to their metamorphosis into lifeless compounds ; his investigations ought to tell us what has taken place and what can take place in the body. It is singular that we find medicinal agencies all dependant on certain matters, which differ in composition ; and if, by the introduction of a substance, certain abnormal conditions are rendered normal, it will be impossible to reject the opinion, that this phenomenon depends on a change in the composition of the constituents of the diseased organism, a change in which the elements of the remedy take a share ; a share similar to that which the vegetable elements of food have taken in the formation of fat, of membranes, of the saliva, of the seminal fluid, &c. Their carbon, hydrogen, or nitrogen, or whatever else belongs to their composition, are derived from the vegetable organism ; and, after all, the action and effects of quinine, morphia, and the vegetable poisons in general, are no hypotheses.

95. Thus, as we may say, in a certain sense, of caffeine or theine and asparagine, &c., as well as of the non-azotised elements of food, that they are food for the liver, since they contain the elements, by the presence of which that organ is enabled to perform its functions, so we may consider these ni-

trogenised compounds, so remarkable for their action
on the brain and on the substance of the organs of
motion, as elements of food for the organs as yet
unknown, which are destined for the metamorphosis
of the constituents of the blood into nervous substance and brain. Such organs there must be in
the animal body, and if, in the diseased state, an abnormal process of production or transformation of
the constituents of cerebral and nervous matter has
been established; if, in the organs intended for this
purpose, the power of forming that matter out of
the constituents of blood, or the power of resisting
an abnormal degree of activity in its decomposition
or transformation, has been diminished; then, in a
chemical sense, there is no objection to the opinion,
that substances of a composition analogous to that
of nervous and cerebral matter, and, consequently,
adapted to form that matter, may be employed, instead of the substances produced from the blood,
either to furnish the necessary resistance, or to restore the normal condition.

96. Some physiologists and chemists have expressed doubts of the peculiar and distinct character
of cerebric acid, a substance which, from its amount
of carbon and hydrogen, and from its external characters, resembles a nitrogenised fatty acid. But
a nitrogenised fat, having an acid character, is,
in fact, no anomaly. Hippuric acid is in many of
its characters very similar to the fatty acids, but is
essentially distinguished from them by containing

nitrogen. The organic constituents of bile resemble the acid resins in physical characters, and yet contain nitrogen. The organic alkalies are intermediate in their physical characters between the fats and resins, and they all contain nitrogen. A nitrogenised fatty acid is as little improbable as the existence of a nitrogenised resin with the characters of a base.

97. An accurate investigation would probably discover differences in the composition of the brain, spinal marrow, and nerves. According to the observations of Valentin, the quality of the cerebral and nervous substance is very rapidly altered from the period of death, and very uncommon precautions would be required for the separation of foreign matters, not properly belonging to the substance of the spinal marrow or. brain. But, however difficult it may appear, the investigation seems yet to be practicable. We know, in the meantime, that all experience is against the notion of a large amount of carbon and hydrogen in the substance of the brain. The absence of nitrogen as an element of the cerebral and nervous matter, appears, at all events, improbable. This substance, moreover, cannot be classed with ordinary fats, because we find the cerebric acid combined with soda, whereas, all fats are compounds of fatty acids with oxide of glycerule. In regard to the phosphorus of the brain, we can only guess as to the form in which the phosphorus exists. Walchner observed re-

cently that bubbles of spontaneously inflammable phosphuretted hydrogen were disengaged from the trough of a spring in Carlsruhe, on the bottom of which fish had putrefied ; and gases containing phosphorus have also been observed among the products of the putrefaction of the brain.*

* The curator of the museum at Geneva gave to M. Leroyer, apothecary, a large quantity of spirit of wine, which had been used for the preservation of fishes, and which he undertook to purify.   He distilled it from a mixture of chloride of calcium and quicklime, and evaporated the residue in the air, over a fire.   As soon as the mass had acquired a certain consistence, and a higher temperature, a prodigious quantity of spontaneously inflammable phosphuretted hydrogen was disengaged.   (Dumas, V. 267.)

PART III.

———•———

THE

PHENOMENA OF MOTION

IN THE

ANIMAL ORGANISM.

PHENOMENA OF MOTION

ANIMAL ORGANISM.

I.

IT might appear an unprofitable task to add one more to the innumerable forms under which the human intellect has viewed the nature and essence of that peculiar cause which must be considered as the ultimate source of the phenomena which characterize vegetable and animal life, were it not that certain conceptions present themselves as necessary deductions from the views on this subject developed in the introduction to the first part of this work. The following pages will be devoted to a more detailed examination of these deductions. It must be admitted here, that all these conclusions will lose their force and significance, if it can be proved that the cause of vital activity has in its manifestations nothing in common with other known causes which produce motion or change of form and structure in matter.

But a comparison of its peculiarities with the modus operandi of these other causes, cannot, at all events, fail to be advantageous, inasmuch as the nature and essence of natural phenomena are recognizable, not by abstraction, but only by comparative observations.

If the vital phenomena be considered as manifestations of a peculiar force, then the effects of this force must be regulated by certain laws, which laws may be investigated; and these laws must be in harmony with the universal laws of resistance and motion, which preserve in their courses the worlds of our own and other systems, and which also determine changes of form and structure in material bodies; altogether independently of the matter in which vital activity appears to reside, or of the form in which vitality is manifested.

The vital force in a living animal tissue appears as a cause of growth in the mass, and of resistance to those external agencies which tend to alter the form, structure, and composition of the substance of the tissue in which the vital energy resides.

This force further manifests itself as a cause of motion and of change in the form and structure of material substances, by the disturbance and abolition of the state of rest in which those chemical forces exist, by which the elements of the compounds conveyed to the living tissues, in the form of food, are held together.

The vital force causes a decomposition of the constituents of food, and destroys the force of attraction which is continually exerted between their molecules; it alters the direction of the chemical forces in such wise, that the elements of the constituents of food arrange themselves in another form, and combine to produce new compounds, either identical in composition with the living tissues, or differing from them; it further changes the direction and force of the attraction of cohesion, destroys the cohesion of the nutritious compounds, and forces the new compounds to assume forms altogether different from those which are the result of the attraction of cohesion when acting freely, that is, without resistance.

The vital force is also manifested as a force of attraction, inasmuch as the new compound produced by the change of form and structure in the food, when it has a composition identical with that of the living tissue, becomes a part of that tissue.

Those newly-formed compounds, whose composition differs from that of the living tissue, are removed from the situation in which they are formed, and, in the shape of certain secretions, being carried to other parts of the body, undergo in contact with these a series of analogous changes.

The vital force is manifested in the form of resistance, inasmuch as by its presence in the living tissues, their elements acquire the power of withstanding the disturbance and change in their form and composition, which external agencies tend to produce; a power which, simply as chemical compounds, they do not possess.

As in the case of other forces, the conception of an unequal intensity of the vital force comprehends not only an unequal capacity for growth in the mass, and an unequal power of overcoming chemical resistance, but also an inequality in the amount of that resistance which the parts or constituents of the living tissue oppose to a change in their form and composition, from the action of new external active causes of change; just as the force of cohesion or of affinity is in direct proportion to the resistance which these forces oppose to any external cause, mechanical or chemical, tending to separate the molecules, or the elements of an existing compound.

The manifestations of the vital force are dependent on a certain form of the tissue in which it resides, as well as on a fixed composition in the substance of the living tissue.

The capacity of growth in a living tissue is determined by the immediate contact with matters adapted to a certain decomposition, or the elements of which are capable of becoming component parts of the tissue in which vitality resides.

The phenomenon of growth, or increase in the mass, presupposes that the acting vital force is more powerful than the resistance which the chemical force opposes to the decomposition or transformation of the elements of the food.

The manifestations of the vital force are dependent on a certain temperature. Neither in a plant nor in an animal do vital phenomena occur when the temperature is lowered to a certain extent.

The phenomena of vitality in a living organism diminish in intensity when heat is abstracted, provided the lost heat be not restored by other causes.

Deprivation of food soon puts a stop to all manifestations of vitality.

The contact of the living tissues with the elements of nutrition is determined in the animal body by a mechanical force produced within the body, which gives to certain organs the power of causing change of place, of producing motion, and of overcoming mechanical resistance.

We may communicate motion to a body at rest by means of a number of forces, very different in their manifestations. Thus, a time-piece may be set in motion by a falling weight (gravitation), or by a bent spring (elasticity). Every kind of motion may be produced by the electric or magnetic force, as well as by chemical attraction; while we cannot say, as long as we only consider the manifestation of these forces in the phenomenon or result produced, which of these various causes of change of place has set the body in motion.

In the animal organism we are acquainted with only one cause of motion; and this is the same cause which determines the growth of living tissues, and gives them the power of resistance to external agencies; it is the vital force.

In order to attain a clear conception of these manifestations of the vital force, so different in form, we must bear in mind, that every known

force is recognized by two conditions of activity,
entirely different in the phenomena they offer to the
attention of the observer.

The force of gravitation inherent in the particles
of a stone, gives to them a continual tendency to
move towards the centre of the earth.

This effect of gravitation becomes inappreciable
to the senses when the stone, for example, rests
upon a table, the particles of which oppose a resistance to the manifestation of its gravitation. The
force of gravity, however, is constantly present, and
manifests itself as a pressure on the supporting
body; but the stone remains at rest; it has no motion. The manifestation of its gravity in the state
of rest we call its weight.

That which prevents the stone from falling is a
resistance produced by the force of attraction, by
which the particles of the wood cohere together; a
mass of water would not prevent the fall of the stone.

If the force which impelled the mass of the stone
towards the centre of the earth were greater than
the force of cohesion in the particles of the wood,
the latter would be overcome; it would be unable
to prevent the fall of the stone.

When we remove the support, and with it the
force which has prevented the manifestation of the
force of gravity, the latter at once appears as the
cause of change of place in the stone, which acquires
motion, or falls. Resistance is invariably the result
of a force in action.

According as the stone is allowed to fall during a longer or shorter time, it acquires properties which it had not while at rest; it acquires, for example, the power of overcoming more feeble or more powerful obstacles, or that of communicating motion to bodies in a state of rest.

If it fall from a certain height it makes a permanent impression on the spot on which it falls; if it fall from a still greater height (during a longer time) it perforates the table; its own motion is communicated to a certain number of the particles of the wood which now fall along with the stone itself. The stone, while at rest, possessed none of these properties.

The velocity of the falling body is always the effect of the moving force, and is, ceteris paribus, proportional to the force of gravitation.

A body, falling freely, acquires at the end of one second a velocity of 30 feet. The same body, if falling on the moon, would acquire in one second only a velocity of $\frac{30}{360}$ths of a foot $=1$ inch, because, in the moon, the intensity of gravitation (the pressure acting on the body, the moving power) is 360 times smaller.

If the pressure continue uniform, the velocity is directly proportional to it; so that, for example, the body falling 360 times slower, will, after 360 seconds, have the same velocity as the other body after one second.

Consequently, the effect is proportional, not to

the moving force alone, nor to the time alone, but to the pressure multiplied into the time, which is called the *momentum of force.*

In two equal masses the velocity expresses the momentum of force. But under the same pressure a body moves more slowly as its mass is greater; a mass twice as great requires, in order to attain in the same time an equal velocity, twice the pressure; or, under the single pressure, it must continue in motion twice as long.

In order, therefore, to have an expression for the whole effect produced, we must multiply the máss into the velocity. This product is called the *amount of motion.*

The amount of motion in a given body must in all cases correspond exactly to the momentum of force.

These two, the amount of motion and the momentum of force, are also called simply *force;* because we suppose that a less pressure acting, for example, during 10 seconds, is equal to a pressure ten times greater, acting only during one second.

The *momentum of motion* in mechanics signifies the effect of a moving force, without reference to the time (velocity) in which it was manifested. If one man, for example, raises 30 lbs. to a height of 100 feet, and a second one 30 lbs. to a height of 200 feet, then the latter has expended twice as much force as the former. A third who raises 60 lbs. to a height of 50 feet, expends no more force than the first did in raising 30 lbs. to the height of

100 feet. The momentum of motion of the first (30×100) is equal to that of the third (60×50), while that of the second (30×200) is twice as great.

Momentum of force and momentum of motion in mechanics are therefore expressions or measures for effects of force, having reference to the velocity attained in a given time, or to a given space; and in this sense they may be applied to the effects of all other causes of motion, or of change in form and structure, however great or however small may be the space or the time in which their effects are displayed to the senses.

Every force, therefore, exhibits itself in matter either in the form of resistance to external causes of motion, or of change in form and structure; or as a moving force when no resistance is opposed to it; or, finally, in overcoming resistance.

One and the same force communicates motion and destroys motion; the former, when its manifestations are opposed by no resistance; the latter, when it puts a stop to the manifestation of some other cause of motion, or of change in form and structure. Equilibrium or rest is that state of activity in which one force or momentum of motion is destroyed by an opposite force or momentum of motion.

We observe both these manifestations of activity in that force which gives to the living tissues their peculiar properties.

The vital force appears as a moving force or cause of motion when it overcomes the chemical forces (cohesion and affinity) which act between the constituents of food, and when it changes the position or place in which their elements occur; it is manifested as a cause of motion in overcoming the chemical attraction of the constituents of food, and is, further, the cause which compels them to combine in a new arrangement, and to assume new forms.

It is plain that a part of the animal body possessed of vitality, which has therefore the power of overcoming resistance, and of giving motion to the elementary particles of the food, by means of the vital force manifested in itself must have a momentum of motion, which is nothing else than the measure of the resulting motion or change in form and structure.

We know that this momentum of motion in the vital force, residing in a living part, may be employed in giving motion to bodies at rest (that is, in causing decomposition, or overcoming resistance), and if the vital force is analogous in its manifestations to other forces, this momentum of motion must be capable of being conveyed or communicated by matters, which in themselves do not destroy its effect by an opposite manifestation of force.

Motion, by whatever cause produced, cannot in itself be annihilated; it may indeed become inappreciable to the senses, but even when arrested by resistance (by the manifestation of an opposite force), its effect is not annihilated. The falling stone, by means of the amount of motion acquired in its descent, produces an effect when it reaches the table. The impression made on the wood, the velocity communicated by its parts to those of the wood,—all this is its effect.

If we transfer the conceptions of motion, equilibrium, and resistance, to the chemical forces, which, in their modus operandi, approach to the vital force infinitely nearer than gravitation does, we know with the utmost certainty, that they are active only in the case of immediate contact. We know also, that the unequal capacity of chemical compounds to offer resistance to external disturbing influences, to those of heat, or of electricity, which tend to separate their particles, as well as their power of overcoming resistance in other compounds (of causing decomposition); that, in a word, the active force in a compound depends on a certain order or arrangement, in which its elementary particles touch each other.

The same elements, united in a different order, when in contact with other compounds, exert a most unequal power of offering or overcoming resistance. In one form the force manifested is available (the body is active, an acid, for example); in another not (the body is indifferent, neutral); in a third form, the momentum of force is opposed to that of the first (the body is active, but a base).

If we alter the arrangement of the elements, we are able to separate the constituents of a compound by means of another active body; while the same elements, united in their original order, would have opposed an invincible resistance to the action of the decomposing agent.

In the same way as two equal inelastic masses, impelled with equal velocity from opposite points, on coming into contact are brought to rest; in the same way, therefore, as two equal and opposite momenta of motion mutually destroy each other; so may the momentum of force in a chemical compound be destroyed in whole or in part by an equal or unequal, and opposite momentum of force in a second compound. But it cannot be annihilated as long as the arrangement of the elementary particles, by which its inherent force was manifested, is not changed.

The chemical force of sulphuric acid is present in sulphate of lime as entire as in oil of vitriol. It is not appreciable by the senses; but if the cause be removed which prevented its manifestation, it appears in its full force in the compound in which it properly resides.

Thus the force of cohesion in a solid may disappear, to the senses, from the action of a chemical force (in solution), or of heat (in fusion), without being in reality annihilated or even weakened. If we remove the opposing force or resistance, the force of cohesion appears unchanged in crystallization.

By means of the electrical force, or that of heat, we can give the most varied directions to the manifestations of chemical force. By these means we can fix, as it were, the order in which the elementary particles shall unite. Let us remove the cause (heat or electricity) which has turned the balance in favour of the weaker attraction in one direction, and the stronger attraction will shew itself continually active in another direction; and if this stronger attraction can overcome the vis inertiæ of the elementary particles, they will unite in a new form, and a new compound of different properties must be the result.

In compounds of this kind, in which, therefore, the free manifestation of the chemical force has been impeded by other forces, a blow, or mechanical friction, or the contact of a substance, the particles of which are in a state of motion (decomposition, transformation), or any external cause, whose activity is added to the stronger attraction of the elementary particles in another direction, may suffice to give the preponderance to this stronger attraction, to overcome the vis inertiæ, to alter the form and structure of the compound, which are the result of foreign causes, and to produce the resolution of the compound into one or more new compounds with altered properties.

Transformations, or as they may be called, phenomena of motion, in compounds of this class, may be effected by means of the free and available chemical force of another chemical compound, and that without its manifestation being enfeebled or arrested by resistance. Thus the equilibrium in the attraction between the elements of cane-sugar is destroyed by contact with a very small quantity of sulphuric acid, and it is converted into grape-sugar. In the same way we see the elements of starch, under the same influence, arrange themselves with those of water in a new form, while the sulphuric acid, which has served to produce these transformations, loses nothing of its chemical character. In regard to other substances on which it acts, it remains as active as before, exactly as if it had exerted no sort of influence on the cane-sugar or starch.

In contradistinction to the manifestations of the so-called mechanical forces, we have recognized in the chemical forces causes of motion and of change in form and structure, without any observable exhaustion of the force by which these phenomena are produced; but the origin of the continued manifestation of activity remains still the same; it is the absence of an opposite force (a resistance) capable of neutralizing it or bringing it into the state of equilibrium.

As the manifestations of chemical forces (the momentum of force in a chemical compound) seem to depend on a certain order in which the elementary particles are united together, so experience tells us, that the vital phenomena are inseparable from matter ; that the manifestations of the vital force in a living part are determined by a certain form of that part, and by a certain arrangement of its elementary particles. If we destroy the form, or alter the composition of the organ, all manifestations of vitality disappear.

There is nothing to prevent us from considering the vital force as a peculiar property, which is possessed by certain material bodies, and becomes sensible when their elementary particles are combined in a certain arrangement or form.

This supposition takes from the vital phenomena nothing of their wonderful peculiarity ; it may therefore be considered as a resting point, from which an investigation into these phenomena, and the laws which regulate them, may be commenced ; exactly as we consider the properties and laws of light to be dependant on a certain luminiferous matter, or ether, which has no further connection with the laws ascertained by investigation.

Considered under this form, the vital force unites in its manifestations all the peculiarities of chemical forces, and of the not less wonderful cause, which we regard as the ultimate origin of electrical phenomena.

The vital force does not act, like the force of gravitation or the magnetic force, at infinite distances, but, like chemical forces, it is active only in the case of immediate contact.   It becomes sensible by means of an aggregation of material particles.

P

A living part acquires, on the above supposition, the capacity of offering and of overcoming resistance, by the combination of its elementary particles in a certain form; and as long as its form and composition are not destroyed by opposing forces, it must retain its energy uninterrupted and unimpaired.

When, by the act of manifestation of this energy in a living part, the elements of the food are made to unite in the same form and structure as the living organ possesses, then these elements acquire the same powers. By this combination, the vital force inherent in them is enabled to manifest itself freely, and may be applied in the same way as that of the previously existing tissue.

If, now, we bear in mind, that all matters which serve as food to living organisms are compounds of two or more elements, which are kept together by certain chemical forces; if we reflect that in the act of manifestation of force in a living tissue, the elements of the food are made to combine in a new order;—it is quite certain that the momentum of force or of motion in the vital force was more powerful than the chemical attraction existing between the elements of the food.*

---

* The hands of a man, who raises with a rope and simple pulley, 30lbs. to the height of 100 feet, pass over a space of 100 feet, while his muscular energy furnishes the equilibrium to a pressure of 30lbs. Were the force which the man could exert not greater than would suffice to keep in equilibrium a pressure of

The chemical force which kept the elements together acted as a resistance, which was overcome by the active vital force.

Had both forces been equal, no kind of sensible effect would have ensued. Had the chemical force been the stronger, the living part would have undergone a change.

If we now suppose that a certain amount of vital force must have been expended in bringing to an equilibrium the chemical force, there must still remain an excess of force, by which the decomposition was effected. This excess constitutes the momentum of force in the living part, by means of which the change was produced ; by means of this excess the part acquires a permanent power of causing further decompositions, and of retaining its condition, form, and structure, in opposition to external agencies.

We may imagine this excess to be removed, and employed in some other form. This would not of itself endanger the existence of the living part, because the opposing forces would be left in equilibrio ; but, by the removal of the excess of force, the part would lose its capacity of growth, its power to cause further decompositions, and its ability to resist external causes of change. If, in this state of equilibrium, oxygen (a chemical agent) should be brought in contact with it, then there would be no

30 lbs., he would be unable to raise the weight to the height mentioned.

resistance to the tendency of the oxygen to combine with some element of the living part, because its power of resistance has been taken away by some other application of its excess of vital force. According to the amount of oxygen brought to it, a certain proportion of the living part would lose its condition of vitality, and take the form of a chemical combination, having a composition different from that of the living tissue. In a word, there would occur a change in the properties of the living compound, or what we have called a change of matter.

If we reflect that the capacity of growth or increase of mass in plants is almost unlimited; that a hundred twigs from a willow tree, if placed in the soil, become a hundred trees; we can hardly entertain a doubt, that with the combination of the elements of the food of the plant so as to form a part of it, a fresh momentum of force is added in the newly formed part to the previously existing momentum in the plant; insomuch, that with the increase of mass, the sum of vital force is augmented.

According to the amount of available vital force, the products formed by its activity from the food are varied. The composition of the buds, of the radical.fibres, of the leaf, of the flower, and of the fruit, are very different one from the other; and the chemical force by which their elements are held together is very different in each of these cases.

Of the non-azotised constituents of plants we

may assert, that no part of the momentum of force
is expended in maintaining their form and structure,
when their elements have once combined in that
order in which they become parts of organs endued
with vitality.

Very different is the character of the azotised
vegetable principles; for, when separated from the
plant, they pass, as is commonly said, spontaneously,
into fermentation and putrefaction. The cause of
this decomposition or transformation of their elements is the chemical action which the oxygen of the
atmosphere exercises on one of their constituents.
Now we know, that as long as the plant exhibits
the phenomena of life, oxygen gas is given off from
its surface; that this oxygen is altogether without
action on the constituents of the living plant, for
which, in other circumstances, it has the strongest
attraction. It is obvious, therefore, that a certain
amount of vital force must be expended, partly to
retain the elements of the complex azotised principles in the form, order, and structure which belong to them; and partly as a means of resistance
against the incessant tendency of the oxygen of the
atmosphere to act on their elements, as well as
against that of the oxygen separated in the organism of the plant by the vital process.

With the increase of these easily altered compounds, in the flower and in the fruit, for example,
the sum of chemical force (the free manifestation
of which, counteracted by an equal measure of vital

force, is employed to furnish resistance) also increases.

The plant increases in mass until the vital force
inherent in it comes into equilibrium with all the
other causes opposed to its manifestation. From
this period, every new cause of disturbance, added
to those previously existing (a change of temperature, for example), deprives it of the power of offering resistance, and it dies down.

In perennial plants (in trees, for example), the
mass of the easily decomposable (azotised) compounds, compared with that of the non-azotised, is
so small, that of the whole sum of force, only a mi-
nimum is expended as resistance. In animals, this
proportion is reversed.

During every period of the life of a plant, the
available vital force (that which is not neutralized
by resistance) is expended only in one form of vital
manifestation, that of growth or increase of mass,
or the overcoming of resistance. No part of this
force is applied to other purposes.

In the animal organism, the vital force exhibits
itself, as in the plant, in the form of the capacity of
growth, and as the means of resistance to external
agencies; but both of these manifestations are confined within certain limits.

We observe in animals, that the conversion of
food into blood, and the contact of the blood with
the living tissues, are determined by a mechanical
force, whose manifestation proceeds from distinct

organs, and is effected by a distinct system of organs, possessing the property of communicating and extending the motion which they receive. We find the power of the animal to change its place and to produce mechanical effects by means of its limbs, dependant on a second similar system of organs or apparatus. Both of these systems of apparatus, as well as the phenomena of motion proceeding from them, are wanting in plants.

In order to form a clear conception of the origin and source of the mechanical motions in the animal body, it may be advantageous to reflect on the modus operandi of other forces, which in their manifestations are most closely allied to the vital force.

When a number of plates of zinc and copper, arranged in a certain order, are brought into contact with an acid, and when the extremities of the apparatus are joined by means of a metallic wire, a chemical action begins at the surface of the plates of zinc, and the wire, in consequence of this action, acquires the most singular and wonderful properties.

The wire appears as the carrier or conductor of a force, which may be conducted and communicated through it in every direction with amazing velocity. It is the conductor or propagator of an uninterrupted series of manifestations of activity.

Such a propagation of motion is inconceivable, if in the wire there were a resistance to be overcome; for every resistance would convert a part of the moving force into a force at rest.

When the wire is divided in the middle, and its continuity interrupted, the propagation of force ceases, and we observe, that in this case the action between the zinc and the acid is immediately stopped.

If the communication be restored, the action which had disappeared reappears with all its original energy.

By means of the force present in the wire, we can produce the most varied effects; we can overcome all kinds of resistance, raise weights, set ships in motion, &c. And, what is still more remarkable, the wire acts as a hollow tube, in which a current of chemical force circulates freely and without hindrance.

Those properties which, when firmly attached to certain bodies, we call the strongest and most energetic affinities, we find, to all appearance, free and uncombined in the wire. We can transport them from the wire to other bodies, and thereby give to them an affinity (a power of entering into combination) which in themselves they do not possess. According to the amount of force circulating in the wire, we are able by means of it to decompose compounds, the elements of which have the strongest attraction for each other. Yet the substance of the wire takes not the smallest share in all these manifestations of force; it is merely the conductor of force.

We observe, further, in this wire, phenomena of

attraction and repulsion, which we must ascribe to
the disturbance of the equilibrium in the electric
or magnetic force; and when this equilibrium is
restored, the restoration is accompanied by the developement of light and heat, its never-failing com-
panions.

All these remarkable phenomena are produced
by the chemical action which the zinc and the acid
exert on each other; they are accompanied by a
change in form and structure, which both undergo.

The acid loses its chemical character; the zinc
enters into combination with it. The manifestations
of force produced in the wire are the immediate
consequence of the change in the properties of the
acid and the metal.

One particle of acid after another loses its peculiar chemical character; and we perceive that in
the same proportion the wire acquires a chemical,
mechanical, galvanic, or magnetic force, whatever
name be given to it. According to the number of
acid particles which in a given time undergo this
change, that is, according to the surface of the zinc,
the wire receives a greater or less amount of these
forces.

The continuance of the current of force depends
on the duration of the chemical action; and the
duration of the latter is most closely connected with
the carrying away, by conduction, of the force.

If we check the propagation of the current of
force, the acid retains its chemical character. If

we employ it to overcome chemical or mechanical resistance, to decompose chemical compounds, or to produce motion, the chemical action continues; that is to say, one particle of acid after another changes its properties.

In the preceding paragraphs we have considered these remarkable phenomena in a form which is independent of the explanations of the schools. Is the force which circulates in the wire the electrical force? Is it chemical affinity? Is it propagated in the conductor like a fluid set in motion, or in the form of a series of momenta of motion, like light and sound, from one particle of the conductor to another? All this we know not, and we shall never know. All the suppositions which may be employed as explanations of the phenomena have not the slightest influence on the truth of these phenomena; for they refer merely to the form in which they are manifested.

On some points, however, there is no doubt; namely, that all the effects which may be produced by the wire are determined by the change of properties in the zinc and in the acid; for the term "chemical action" signifies neither more nor less than the act of change in them; that these effects depend on the presence of a conductor, of a substance which propagates in all directions, where it is not neutralized by resistance, the force or momentum produced; that this force becomes a momentum of motion, by means of which we can pro-

duce mechanical effects, and which, when transferred
to other bodies, communicates to them all those properties, the ultimate cause of which is the chemical
force itself; for these bodies acquire the power of
causing decompositions and combinations, such as,
without a supply of force through the conductor,
they could not effect.

If we employ these well-known facts as means
to assist us in investigating the ultimate cause of
the mechanical effects in the animal organism, observation teaches us, that the motion of the blood
and of the other animal fluids proceeds from distinct
organs, which, as in the case of the heart and intestines, do not generate the moving power in themselves, but receive it from other quarters.

We know with certainty that the nerves are the
conductors and propagators of mechanical effects;
we know, that by means of them motion is propagated in all directions.   For each motion we recognize a separate nerve, a peculiar conductor, with
the conducting power of which, or with its interruption, the propagation of motion is affected or destroyed.

By means of the nerves all the parts of the body,
all the limbs, receive the moving force which is indispensable to their functions, to change of place,
to the production of mechanical effects.   Where
nerves are not found, motion does not occur.   The
excess of force generated in one place is conducted
to other parts by the nerves.   The force which one

organ cannot produce in itself is conveyed to it
from other quarters; and the vital force which is
wanting to it, in order to furnish resistance to external causes of disturbance, it receives in the form
of excess from another organ, an excess which that
organ cannot consume in itself.

We observe further, that the voluntary and involuntary motions, in other words, all mechanical
effects in the animal organism, are accompanied by,
nay, are dependant on, a peculiar change of form
and structure in the substance of certain living
parts, the increase or diminution of which change
stands in the very closest relation to the measure of
motion, or the amount of force consumed in the
motions performed.

As an immediate effect of the manifestation of
mechanical force, we see, that a part of the muscular substance loses its vital properties, its character of life; that this portion separates from the
living part, and loses its capacity of growth and its
power of resistance. We find that this change of
properties is accompanied by the entrance of a
foreign body (oxygen) into the composition of the
muscular fibre (just as the acid lost its chemical
character by combining with zinc); and all experience proves, that this conversion of living muscular
fibre into compounds destitute of vitality is accelerated or retarded according to the amount of force
employed to produce motion. Nay, it may safely
be affirmed, that they are mutually proportional;

that a rapid transformation of muscular fibre, or, as it may be called, a rapid change of matter, determines a greater amount of mechanical force; and conversely, that a greater amount of mechanical motion (of mechanical force expended in motion) determines a more rapid change of matter.

From this decided relation between the change of matter in the animal body and the force consumed in mechanical motion, no other conclusion can be drawn but this, that the active or available vital force in certain living parts is the cause of the mechanical phenomena in the animal organism.

The moving force certainly proceeds from living parts; these parts possessed a momentum of force or of motion, which they lost in proportion as other parts acquired a momentum of force or of motion; they lose their capacity of growth, and their power to resist external causes of change. It is obvious that the ultimate cause, the vital force, from which they acquired those properties, has served for the production of mechanical force, that is, has been expended in the shape of motion.

How, indeed, could we conceive that a living part should lose the condition of life, should become incapable of resisting the action of the oxygen conveyed to it by the arterial blood, and should be deprived of the power to overcome chemical resistance, unless the momentum of the vital force, which had given to it all these properties, had been expended for other purposes?

By the power of the conductors, the nerves, to
propagate the momentum of force in a living part,
or the effect which the active vital force inherent in
the part produces.on all the surrounding parts, in
all directions where the force, or rather its momentum of motion, is consumed without resistance
(for without motion no change of matter occurs,
and when motion has begun, there is no longer resistance), an equilibrium is obviously established in
the living part, between the chemical forces and
the remaining vital force ; which equilibrium would
not have occurred had not vital force been expended
in producing mechanical motion.

In this state, any external cause capable of exerting an influence on the form, structure, and composition of the organ meets with no further resistance. If oxygen were not conveyed to it, the
organ would maintain its condition, but without any
manifestation of vitality. It is only with the commencement of chemical action that the change of
matter, that is, the separation of a part of the organ
in the form of lifeless compounds, begins.

The change of matter, the manifestation of mechanical force, and the absorption of oxygen, are, in
the animal body, so closely connected with each
other, that we may consider the amount of motion,
and the quantity of living tissue transformed, as
proportional to the quantity of oxygen inspired and
consumed in a given time by the animal. For a
certain amount of motion, for a certain proportion

of vital force consumed as mechanical force, an equivalent of chemical force is manifested; that is, an equivalent of oxygen enters into combination with the substance of the organ which has lost the vital force; and a corresponding proportion of the substance of the organ is separated from the living tissue in the shape of an oxidised compound.

All those parts of the body which nature has destined to effect the change of matter, that is, to the production of mechanical force, are penetrated in all directions by a multitude of the most minute tubes or vessels, in which a current of oxygen continually circulates, in the form of arterial blood. To the above-mentioned separation of part of the elements of these parts, in other words, to the disturbance of their equilibrium, this oxygen is absolutely essential.

As long as the vital force of these parts is not conducted away and applied to other purposes, the oxygen of the arterial blood has not the slightest effect on the substance of the organized parts; and in all cases, only so much oxygen is taken up as corresponds to the conducting power, and, consequently, to the mechanical effects produced.

The oxygen of the atmosphere is the proper, active, external cause of the waste of matter in the animal body; it acts like a force which disturbs and tends to destroy the manifestation of the vital force at every moment. But its effect as a chemical agent, the disturbance proceeding from it, is

held in equilibrium by the vital force, which is free
and available in the living tissue, or is annihilated
by a chemical agency opposed to that of oxygen,
the manifestation of which must be considered as
dependant on the vital force.

In chemical language, to annihilate the chemical
action of oxygen, means, to present to it substances,
or parts of organs, which are capable of combining
with it.

The action of oxygen (affinity) is either neutralized by means of the elements of organized parts,
which combine with it (after the free vital force has
been conducted away), or else the organ presents to
it the products of other organs, or certain matters
formed from the elements of the food, by the vital
activity of certain systems of apparatus.

It is only the muscular system which, in this
sense, produces in itself a resistance to the chemical action of oxygen, and neutralizes it completely.

The substance of cellular tissue, of membranes,
and of the skin, the minutest particles of which are
not in immediate contact with arterial blood (with
oxygen), are not destined to undergo this change of
matter. Whatever changes they may undergo in
the vital process, affect, in all cases, only their
surface.

The gelatinous tissues, mucous membranes, tendons, &c., are not designed to produce mechanical
force ; they contain in their substance no con-

ductors of mechanical effects. But the muscular
system is interwoven with innumerable nerves.
The substance of the uterus is in no respect different in chemical composition from the other muscles; but it is not adapted to the change of matter,
to the production of force, and contains no organs
for conducting away the moving power. Cellular
tissue, gelatinous membranes, and mucous membranes, are far from being destitute of the power of
combining with oxygen, when moisture is present;
we know that, when moist, they cannot be brought
in contact with oxygen without undergoing a progressive alteration. But one surface of the intestines and the cells of the lungs are constantly in
contact with oxygen; and it is obvious that they
must be as rapidly altered by the chemical action of
the oxygen in the body as out of it, were it not
that there exists in the organism itself a source of
resistance, which completely neutralizes the action
of the oxygen. Among the means by which this
resistance is furnished we may include all substances which are capable of combining with oxygen, or acquire that property under the influence
of the vital force, and which surpass the tissues
above mentioned in their power of neutralizing its
chemical action.

All those constituents of the body which, in
themselves, do not possess, in the form of vital
force, the power of resisting the action of oxygen,
must be far better adapted for the purpose of com-

bining with, and neutralizing it, than those tissues
which are under the influence of the vital force,
although only through the nerves. In this point of
view, we cannot fail to perceive the importance of
the bile in regard to the substance of the intestines,
and that of the pulmonary cells, as well as that of
fat, of mucus, and of the secretions generally.

When the membranes are compelled from their
own substance to furnish resistance to the action
of the oxygen, that is, when there is a deficiency of
the substances destined by nature for their protection, they must, since their renewal is confined
within narrow limits, yield to the chemical action.
The lungs and intestines will always simultaneously
suffer abnormal changes.

From the change of matter itself, from the metamorphosis of the living muscular tissue, these organs
receive the means of resistance to the action of oxygen which are indispensable to their preservation.
According to the rapidity of this process, the quantity of bile secreted increases ; while that of the fat
present in the body diminishes in the same proportion.

For carrying on the involuntary motions in the
animal body, a certain amount of vital force is expended at every moment of its existence ; and, consequently, an incessant change of matter goes on ;
but the amount of living tissue, which, in consequence of this form of consumption of vital force,
loses its condition of life and its capacity of growth,

is confined within narrow limits. It is directly proportional to the force required for these involuntary motions.

Now, although we may suppose that the living muscular tissue, with a sufficient supply of food, never loses its capacity of growth; that this form of vital manifestation is continually effective; this cannot apply to those parts of the body whose available vital force has been expended in producing mechanical effects. For the waste of matter, in consequence of motion and laborious exertion, is extremely various in different individuals.

If we reflect, that the slightest motion of a finger consumes force; that in consequence of the force expended, a corresponding portion of muscle diminishes in volume; it is obvious, that an *equilibrium between supply and waste of matter* (in living tissues) can only occur when the portion separated or expelled in a lifeless form is, at the same instant in which it loses its vital condition, restored in another part.

The capacity of growth or increase in mass depends on the momentum of force belonging to each part; and must be capable of continued manifestation (if there be a sufficient supply of nourishment), as long as it does not lose this momentum, by expending it, for example, in producing motion.

In all circumstances, the growth itself is restricted to the time; that is to say, it cannot be unlimited in a limited time.

Q 2

A living part cannot increase in volume at the same moment in which a portion of it loses the vital condition, and is expelled from the organ in the form of a lifeless compound ; on the contrary, its volume must diminish.

The continued application of the momentum of force in living tissues to mechanical effects determines, therefore, a continued separation of matter ; and only from the period at which the cause of waste ceases to operate, can the capacity of growth be manifested.

Now, since, in different individuals, according to the amount of force consumed in producing voluntary mechanical effects, unequal quantities of living tissue are wasted, there must occur, in every individual, unless the phenomena of motion are to cease entirely, a condition in which all voluntary motions are completely checked, in which, therefore, these occasion no waste. This condition is called *sleep*.

The growth of one part, which is not deprived of its vital force, cannot be in the slightest degree affected by the consumption of the vital force of another part in producing motion. The one may increase in volume, while the other diminishes ; and the waste in one can neither increase nor diminish the supply in the other.

Now, since the consumption of force for the involuntary motions continues in sleep, it is plain that a waste of matter also continues in that state ; and if the original equilibrium is to be restored, we must suppose that, during sleep, an amount of force is accumulated in the form of living tissue, exactly equal to that which was consumed in voluntary and involuntary motion during the preceding waking period.

If the equilibrium between waste and supply of matter be in the least degree disturbed, this is instantly seen in the different amount of force available for mechanical purposes.

It is further obvious, that if there should occur a disproportion between the conducting power of the nerves of voluntary and involuntary motion, a difference in the phenomena of motion themselves will be perceptible, in the same proportion as the one or the other is capable of propagating the momentum of force, generated by the change of matter. As the motions of the circulating system and of the intestines increase, the power of producing mechanical effects in the limbs must diminish in the same proportion (as in wasting fevers); and if, in a given time, more vital force has been consumed for mechanical purposes (labour, running, dancing, &c.) than is properly available for the voluntary and involuntary motions; if force be expended more rapidly than the change of matter can be effected in the same time; then a part of that force which is necessary for the involuntary motions must be expended in restoring the excess of force consumed in voluntary motion. The motions of the heart and of the intestines, in this case, will be retarded, or will entirely cease.

From the unequal degree of conducting power in the nerves, we must deduce those conditions which are termed paralysis, syncope, and spasm. Paralysis of the nerves of voluntary motion may exist without emaciation; but frequently recurring attacks of epilepsy (in which vital force is rapidly wasted in producing mechanical effects) are always accompanied by remarkably rapid emaciation.

It ought to excite the highest admiration when we consider with what infinite wisdom the Creator has divided the means by which animals and plants are qualified for their functions, for their peculiar vital manifestations.

The living part of a plant acquires the whole force and direction of its vital energy from the absence of all conductors of force. By this means the leaf is enabled to overcome the strongest chemical attractions, to decompose carbonic acid, and to assimilate the elements of its nourishment.

In the flower alone does a process similar to the change of matter in the animal body occur. There, phenomena of motion appear; but the mechanical effects are not propagated to a distance, owing to the absence of conductors of force.

The same vital force which we recognize in the plant as an almost unlimited capacity of growth, is converted in the animal body into moving power (into a current of vital force); and a most

wonderful and wise economy has destined for the
nourishment of the animal only such compounds as
have a composition identical with that of the organs
which generate force, that is, with the muscular
tissue. The expenditure of force which the living
parts of animals require, in order to reproduce
themselves from the blood; the resistance of the
chemical force which has to be overcome in the
azotised constituents of food by the vital agency of
the organs destined to convert them into blood;
these are as nothing compared to the force with
which the elements of carbonic acid are held together. A certain amount of force would necessarily be prevented from assuming the form of moving power, if it were to be expended in overcoming
chemical resistance; for the momentum of motion
of the vital force is diminished by all obstacles. But
the conversion of the constituents of blood into muscular fibre (into an organ which generates force) is
only a change of form. Both have the same composition; blood is fluid, muscular fibre is solid blood.
We may even suppose that this change takes place
without any expenditure of vital force; for the mere
passage of a fluid body into the solid state requires
no manifestation of force, but only the removal of
obstacles, which oppose that force (cohesion), which
determines the form of matter, in its manifestations.

In what form or in what manner the vital force
produces mechanical effects in the animal body is
altogether unknown, and is as little to be ascer-

tained by experiment as the connection of chemical
action with the phenomena of motion which we can
produce with the galvanic battery. All the explanations which have been attempted are only repre-
sentations of the phenomenon; they are, more or
less, exact descriptions and comparisons of known
phenomena with these, whose cause is unknown.
In this respect we are like an ignorant man, to
whom the rise and fall of an iron rod in a cylinder,
in which the eye can perceive nothing, and its connection with the turning and motion of a thousand
wheels at a distance from the piston-rod, appear
incomprehensible.

We know not how a certain something, invisible
and imponderable in itself (heat), gives to certain
bodies the power of exerting an enormous pressure
on surrounding objects; we know not even how this
something itself is produced when we burn wood or
coals.

So is it with the vital force, and with the phenomena exhibited by living bodies. The cause of
these phenomena is not chemical force; it is not
electricity, nor magnetism; it is a force which has
certain properties in common with all causes of
motion and of change in form and structure in material substances. It is a peculiar force, because it exhibits manifestations which are found in no other
known force.

II.

In the living plant, the intensity of the vital force far exceeds that of the chemical action of oxygen.

We know, with the utmost certainty, that, by the influence of the vital force, oxygen is separated from elements to which it has the strongest affinity; that it is given out in the gaseous form, without exerting the slightest action on the juices of the plant.

How powerful, indeed, must the resistance appear which the vital force supplies to leaves charged with oil of turpentine or tannic acid, when we consider the affinity of oxygen for these compounds!

This intensity of action or of resistance the plant obtains by means of the sun's light; the effect of which in chemical actions may be, and is, compared to that of a very high temperature (a moderate red heat).

During the night an opposite process goes on in the plant; we see then that the constituents of the leaves and green parts combine with the oxygen of the air, a property which in daylight they did not possess.

From these facts we can draw no other conclusion but this: that the intensity of the vital force diminishes with the abstraction of light; that with the approach of night a state of equilibrium is established, and that in complete darkness all those con-

stituents of plants which, during the day, possessed
the power of separating oxygen from chemical combinations, and of resisting its action, lose this power
completely.

A precisely similar phenomenon is observed in
animals.

The living animal body exhibits its peculiar manifestations of vitality only at certain temperatures.
When exposed to a certain degree of cold, these
vital phenomena entirely cease.

The abstraction of heat must, therefore, be viewed
as quite equivalent to a diminution of the vital
energy ; the resistance opposed by the vital force to
external causes of disturbance must diminish, in
certain temperatures, in the same ratio in which the
tendency of the elements of the body to combine
with the oxygen of the air increases.

By the combination of oxygen with the constituents of the metamorphosed tissues, the temperature necessary to the manifestations of vitality is
produced in the carnivora. In the herbivora, again,
a certain amount of heat is developed by means of
those elements of their non-azotised food which
have the property of combining with oxygen.

It is obvious that the temperature of an animal
body cannot change, if the amount of inspired oxygen increases in the same ratio as the loss of heat
by external cooling.

Two individuals, carnivora, of equal weight, exposed to unequal degrees of cold, lose, in a given

time, by external cooling, unequal quantities of heat. Experience teaches, that if their peculiar temperature and their original weight are to remain unaltered, they require unequal quantities of food ; more in the lower temperature than in the higher. The circumstance that the original weight remains the same, with unequal quantities of food, obviously presupposes, that in the same time a quantity of oxygen proportional to the temperature has been absorbed ; more in the lower than in the higher temperature.

We find that the weight of both individuals, at the end of 24 hours, is equal to the original weight. But we have assumed that their food is converted into blood ; that the blood has served for nutrition ; and it is plain, that when the original weight has been restored, a quantity of the constituents of the body, equal in weight to those of the food, has lost its condition of life, and has been expelled in combination with oxygen.

The one individual, which, being exposed to the lower temperature, consumed more food, has also absorbed more oxygen ; a greater quantity of the constituents of its body has been separated in combination with oxygen ; and, in consequence of this combination with oxygen, a greater amount of heat has been liberated, by which means the heat abstracted has been restored, and the proper temperature of the body kept up.

Consequently, by the abstraction of heat, provided

there be a full supply of food and free access of
oxygen, the change of matter must be accelerated ;
and, along with the augmented transformation, in a
given time, of living tissues, a greater amount of
vital force must be rendered available for mechanical purposes.

With the external cooling, the respiratory motions become stronger; in a lower temperature more
oxygen is conveyed to the blood; the waste of
matter increases, and if the supply be not kept in
equilibrium with this waste, by means of food, the
temperature of the body gradually sinks.

But, in a given time, an unlimited supply of
oxygen cannot be introduced into the body; only a
certain amount of living tissue can lose the state of
life, and only a limited amount of vital force can be
manifested in mechanical phenomena. It is only,
therefore, when the cooling, the generation of force,
and the absorption of oxygen are in equilibrium
together, that the temperature of the body can remain unchanged. If the loss of heat by cooling go
beyond a certain point, the vital phenomena diminish in the same ratio ; for the temperature falls,
and the temperature must be considered as a uniform condition of their manifestation.

Now experience teaches, that when the temperature of the body sinks, the power of the limbs to
produce mechanical effects (or the force necessary
to the voluntary motions) is also diminished. The
condition of sleep ensues, and at last even the invo-

luntary motions (those of the heart and intestines, for example) cease, and apparent death or syncope supervenes.

It is obvious that the cause of the generation of force, namely the change of matter, is diminished, because, with the abstraction of heat, as in the plant by abstraction of light, the intensity of the vital force diminishes. It is also obvious that the momentum of force in a living part depends on its proper temperature; exactly as the effect of a falling body stands in a fixed relation to certain other conditions; for example, to the velocity attained in falling.

When the temperature sinks, the vital energy diminishes; when it again rises, the momentum of force in the living parts appears once more in all its original intensity.

The production of force for mechanical purposes, and the temperature of the body, must, consequently, bear a fixed relation to the amount of oxygen which can be absorbed in a given time by the animal body.

The quantities of oxygen which a whale and a carrier's horse can inspire in a given time are very unequal. The temperature, as well as the quantity of oxygen, is much greater in the horse.

The force exerted by a whale, when struck with the harpoon, his body being supported by the surrounding medium, and the force exerted by a carrier's horse, which carries its own weight and a heavy burden for eight or ten hours, must both bear the same ratio to the oxygen consumed. If we

take into consideration the time during which the
force is manifested, it is obvious that the amount of
force developed by the horse is far greater than in
the case of the whale.

In climbing high mountains, where, in consequence of the respiration of a highly rarefied atmosphere, much less oxygen is conveyed to the blood,
in equal times, than in valleys or at the level of the
sea, the change of matter diminishes in the same
ratio, and with it the amount of force available for
mechanical purposes. For the most part, drowsiness
and want of force for mechanical exertions come
on ; after twenty or thirty steps, fatigue compels us
to a fresh accumulation of force by means of rest
(absorption of oxygen without waste of force in
voluntary motions).

By the absorption of oxygen into the substance
of living tissues, these lose their condition of life,
and are separated as lifeless, unorganised compounds ; but the whole of the inspired oxygen is not
applied to these transformations : the greater part
serves to convert into gas and vapour all matters
which no longer belong to the organism ; and, as
formerly mentioned, the combination of the elements of such compounds with the oxygen produces
the temperature proper to the animal organism.

The production of heat and the change of matter
are closely related to each other ; but although heat
can be produced in the body without any change of
matter in living tissues, yet the change of matter

cannot be supposed to take place without the co-operation of oxygen.

According to all the observations hitherto made, neither the expired air, nor the perspiration, nor the urine, contains any trace of alcohol, after indulgence in spirituous liquors; and there can be no doubt that the elements of alcohol combine with oxygen in the body ; that its carbon and hydrogen are given off as carbonic acid and water.

The oxygen which has accomplished this change must have been taken from the arterial blood; for we know of no channel, save the circulation of the blood, by which oxygen can penetrate into the interior of the body.

Owing to its volatility, and the ease with which its vapour permeates animal membranes and tissues, alcohol can spread throughout the body in all directions.

If the power of the elements of alcohol to combine with oxygen were not greater than that of the compounds formed by the change of matter, or that of the substance of living tissues, they (the elements of alcohol) could not combine with oxygen in the body.

It is, consequently, obvious, that by the use of alcohol a limit must rapidly be put to the change of matter in certain parts of the body. The oxygen of the arterial blood, which, in the absence of alcohol, would have combined with the matter of the tissues, or with that formed by the metamorphosis of these tissues, now combines with the elements of alcohol. The arterial blood becomes venous, without the substance of the muscles having taken any share in the transformation.

Now we observe, that the developement of heat in the body, after the use of wine, increases rather than diminishes, without the manifestation of a corresponding amount of mechanical force.

A moderate quantity of wine, in women and children unaccustomed to its use, produces, on the contrary, a diminution of the force necessary for voluntary motions. Weariness, feebleness in the limbs, and drowsiness, plainly shew that the force available for mechanical purposes, in other words, the change of matter, has been diminished.

A diminution of the conducting power of the nerves of voluntary motion may doubtless take a certain share in producing these symptoms; but this must be altogether without influence on the sum of available force.

What the conductors of voluntary motion cannot carry away for effects of force, must be taken up by the nerves of involuntary motion, and conveyed to the heart, lungs, and intestines. In this case, the circulation will appear accelerated at the expense of the force available for voluntary motion; but, as was before remarked, without the production of a greater amount of mechanical force by the process of oxidation of the alcohol.

Finally, we observe, in hybernating animals, that,

during their winter sleep, the capacity of increase in
mass (one of the chief manifestations of the vital
force), owing to the absence of food, is entirely suppressed. In several, apparent death occurs in consequence of the low temperature and of the diminution of vital energy thus produced ; in others, the
involuntary motions continue, and the animal pre-
serves a temperature independent of the surrounding temperature. The respirations go on ; oxygen,
the condition which determines the production of
heat and of force, is absorbed now as well as in the
former state of the animal ; and previous to the
winter sleep, we find all those parts of their body,
which in themselves are unable to furnish resistance
to the action of the oxygen, and which, like the
intestines and membranes, are not destined for the
change of matter, covered with fat; that is, sur-
rounded by a substance which supplies the want of
resistance.

If we now suppose, that the oxygen absorbed
during the winter sleep combines, not with the èle-
ments of living tissues, but with those of the fat,
then the living part, although a certain momentum
of motion be expended in keeping up the circulation, will not be separated and expelled from the
body.

With the return of the higher temperature, the
capacity of growth increases in the same ratio, and
the motion of the blood increases with the absorption of oxygen. Many of these animals become

R

emaciated during the winter sleep, others not till after awaking from it.

In hybernating animals the active force of the living parts is exclusively devoted, during hybernation, to the support of the involuntary motions. The expenditure of force in voluntary motion is entirely suppressed.

In contradistinction to these phenomena, we know that, in the case of excess of motion and exertion, the active force in living parts may be exclusively and entirely consumed in producing voluntary mechanical effects; in suchwise that no force shall remain available for the involuntary motions. A stag may be hunted to death; but this cannot occur without the metamorphosis of all the living parts of its muscular system, and its flesh becomes uneatable. The condition of metamorphosis into which it has been brought by an enormous consumption both of force and of oxygen, continues when all phenomena of motion have ceased. In the living tissues, all the resistance offered by the vital force to external agencies of change is entirely destroyed.

But however closely the conditions of the production of heat and of force may seem to be connected together, with reference to mechanical effects, yet the disengagement of heat can in no way be considered as in itself the only cause of these effects.

All experience proves, that there is, in the organism, only one source of mechanical power; and this source is the conversion of living parts into lifeless, amorphous compounds.

Proceeding from this truth, which is independent of all theory, animal life may be viewed as determined by the mutual action of opposed forces; of which one class must be considered as *causes of increase* (of supply of matter), and the other as *causes of diminution* (of waste of matter).

The increase of mass is effected in living parts by the vital force; the manifestation of this power is dependant on heat; that is, on a certain temperature peculiar to each specific organism.

The cause of waste of matter is the chemical action of oxygen; and its manifestation is dependant on the abstraction of heat as well as on the expenditure of the vital force for mechanical purposes.

The act of waste of matter is called the change of matter; it occurs in consequence of the absorption of oxygen into the substance of living parts. This absorption of oxygen occurs only when the resistance which the vital force of living parts opposes to the chemical action of the oxygen is weaker than that chemical action; and this weaker resistance is determined by the abstraction of heat, or by the expenditure in mechanical motions of the available force of living parts.

By the combination of the oxygen introduced in the arterial blood with such constituents of the body as offer no resistance to its action, the temperature necessary for the manifestation of vital activity is produced.

From the relations between the consumption of oxygen on the óne hand, and the change of matter and developement of heat on the other, the following general rules may be deduced.

For every proportion of oxygen which enters into combination in the body, a corresponding proportion of heat must be generated.

The sum of force available for mechanical purposes must be equal to the sum of the vital forces of all tissues adapted to the change of matter.

If, in equal times, unequal quantities of oxygen are consumed, the result is obvious, in an unequal amount of heat liberated, and of mechanical force.

When unequal amounts of mechanical force are expended, this determines the absorption of corresponding and unequal quantities of oxygen.

For the conversion of living tissues into lifeless compounds, and for the combination of oxygen with such constituents of the body as have an affinity for it, *time is required.*

In a given time, only a limited amount of mechanical force can be manifested, and only a limited amount of heat can be liberated.

That which is expended, in mechanical effects, in the shape of velocity, is lost in time; that is to say, the more rapid the motions are, the sooner or the more quickly is the force exhausted.

The sum of the mechanical force produced in a

given time is equal to the sum of force necessary, during the same time, to produce the voluntary and involuntary motions; that is, all the force which the heart, intestines, &c., require for their motions is lost to the voluntary motions.

The amount of azotised food necessary to restore the equilibrium between waste and supply is directly proportional to the amount of tissues metamorphosed.

The amount of living matter, which in the body loses the condition of life, is, in equal temperatures, directly proportional to the mechanical effects produced in a given time.

The amount of tissue metamorphosed in a given time may be measured by the quantity of nitrogen in the urine.

The sum of the mechanical effects produced in two individuals, in the same temperature, is proportional to the amount of nitrogen in their urine; whether the mechanical force has been employed in voluntary or involuntary motions, whether it has been consumed by the limbs or by the heart and other viscera.

That condition of the body which is called *health* includes the conception of an equilibrium among all the causes of waste and of supply; and thus animal life is recognized as the mutual action of both; and appears as an alternating destruction and restoration of the state of equilibrium.

In regard to its absolute amount, the waste and

supply of matter is, in the different periods of life,
unequal; but, in the state of health, the available
vital force must always be considered as a constant
quantity, corresponding to the sum of living particles.

Growth, or the increase of mass, stands, at every
age, in a fixed relation to the amount of vital force
consumed as moving power.

The vital force, which is expended for mechanical
purposes, is subtracted from the sum of the force
available for the purpose of increase of mass.

The active force, which is consumed in the body
in overcoming resistance (in causing increase of
mass), cannot, at the same time, be employed to
produce mechanical effects.

Hence it follows necessarily, that when, as in
childhood, the supply exceeds the waste of matter,
the mechanical effects produced must be less in the
same proportion.

With the increase of mechanical effects produced,
the capacity of increase of mass or of the supply of
waste in living tissues must diminish in the same
proportion.

A perfect balance between the consumption of
vital force for supply of matter and that for mechanical effects occurs, therefore, only in the adult
state. It is at once recognized in the complete
supply of the matter consumed. In old age more
is wasted; in childhood more is supplied than
wasted.

The force available for mechanical purposes in an adult man is reckoned, in mechanics, equal to $\frac{1}{5}$th of his own weight, which he can move during eight hours, with a velocity of five feet in two seconds.

If the weight of a man be 150 lbs., his force is equal to a weight of 30 lbs. carried by him to a distance of 72,000 feet. For every second his momentum of force is $= 30 \times 2\cdot 5 = 75$ lbs.; and for the whole day's work, his momentum of motion is $= 30 \times 72,000 = 216,000$.

By the restoration of the original weight of his body, the man collects again a sum of force which allows him, next day, to produce, without exhaustion, the same amount of mechanical effects.

This supply of force is furnished in a seven hours' sleep.

In manufactories of rolled iron it frequently happens, that the pressure of the engine, going at its ordinary rate, is not sufficient to force a rod of iron of a certain thickness to pass below the cylinders. The workman, in this case, allows the whole force of the steam to act on the revolving wheel, and not until this has acquired a great velocity does he bring the rod under the rollers; when it is instantly flattened with great ease into a plate, while the wheel gradually loses the velocity it had acquired. What the wheel gained in velocity, the roller gained in force; by this process force was obviously collected, accumulated in the velocity; but in this sense force does not accumulate in the living organism.

The restoration of force is effected, in the animal body, by the transformation of the separated parts, destined for the production of force, and by the expenditure of the active vital force in causing *formation of new parts;* and, with the restoration of the separated or effete parts, the organism recovers a force equal to that which has been expended.

It is plain, that the vital force manifested, during sleep, in the formation of new parts must be equal to the whole sum of the moving power expended in the waking state in all mechanical effects whatever, *plus* a certain amount of force, which is required for carrying on those involuntary motions which continue during sleep.

From day to day, the labouring man, with sufficient food, recovers, in seven hours' sleep, the whole sum of force; and without reckoning the force necessary for the involuntary motions which may be considered equal in all men, we may assume, that the mechanical force available for work is directly proportional to the number of hours of sleep.

The adult man sleeps 7 hours, and wakes 17 hours; consequently, *if the equilibrium be restored* in 24 hours, the mechanical effects produced in 17 hours must be equal to the effects produced during 7 hours in the shape of formation of new parts.

An old man sleeps only $3\frac{1}{2}$ hours; and if every thing else be supposed the same as in the case of the adult, he will be able, at all events, to produce half of the mechanical effects produced by an adult of equal weight ; that is, he will be able to carry only 15 lbs. instead of 30 to the same distance.

The infant at the breast sleeps 20 hours and wakes only four ; the active force consumed in formation of new parts is, in this case, to that consumed in mechanical effects (in motion of the limbs), as 20 to 4 ; but his limbs possess no momentum of force, for he cannot yet support his own body. If we assume, that the aged man and infant consume in mechanical effects a quantity of force corresponding to the proportion available in the adult, then the mechanical effects are proportional to the number of waking hours, the formation of new parts to the number of hours of sleep, and we shall have :

Force expended in mechanical effects.		*Force expended in formation of new parts.*
In the adult 17	:	7
In the infant 4	:	20
In the old man 20	: ·	4

In the adult, a perfect equilibrium takes place between waste and supply ; in the old man and in the infant, waste and supply are not in equilibrium. If we make the consumption of force in the 17 waking hours equal to that required for the restoration of the equilibrium during sleep = 100 = 17 waking hours, = 7 hours of sleep, we obtain the following proportions. The mechanical effects are to those in the shape of formation of new parts :

In the adult man = 100 : 100
In the infant ... = 25 : 250
In the old man... = 125 : 50

Or the increase of mass to the diminution by
waste:

> In the adult man = 100 : 100
> In the infant ... = 100 : 10
> In the old man... = 100 : 250

It is consequently clear, that if the old man
performs an amount of work proportional to the
sleeping hours of the adult, the waste will be greater
than the supply; that is, his body will rapidly decrease in weight, if he carry 15 lbs. to the distance
of 72,000 feet with a velocity of $2\frac{1}{2}$ feet in the
second; but he will be able, without injury, to
carry 6 lbs. to the same distance.

In the infant the increase is to the decrease as
10 to 1, and consequently, if we in his case increase
the expenditure of force in mechanical effects to ten
times its proper amount, there will thus be established only an equilibrium between waste and supply. The child, indeed, will not grow; but neither
will it lose weight.

If, in the adult man, the consumption of force
for mechanical purposes in 24 hours be augmented
beyond the amount restorable in seven hours of
sleep, then, if the equilibrium is to be restored, less
force, in the same proportion, must be expended in
mechanical effects in the next 24 hours. If this be
not done, the mass of the body decreases, and the
state characteristic of old age more or less decidedly
supervenes.

With every hour of sleep the sum of available

force increases in the old man, or approaches the
state of equilibrium between waste and supply
which exists in the adult.

It is further evident, that if a part of the force
which is available for mechanical purposes, without
disturbing the equilibrium, should not be consumed
in moving the limbs, in raising weights, or in other
labour, it will be available for involuntary motions.
If the motion of the heart, of the fluids, and of the
intestines (the circulation of the blood and digestion), are accelerated in proportion to the amount
of force not consumed in voluntary motions, the
weight of the body will neither increase nor diminish
in 24 hours. The body, therefore, can only increase
in mass, if the force accumulated during sleep, and
available for mechanical purposes, is employed neither for voluntary nor for involuntary motions.

The numerical values above given for the expenditure of force in the human body refer, as has been
expressly stated, only to a given, uniform temperature. In a different temperature, and with deficient nourishment, all these proportions must be
changed.

If we surround a part of the body with ice or
snow, while other parts are left in the natural state,
there occurs, more or less quickly, in consequence of
the loss of heat, an accelerated change of matter in
the cooled part.

The resistance of the living tissues to the action
of oxygen is weaker at the cooled part than in the

other parts ; and this, in its effects, is equivalent to an increase of resistance in these other parts.

The momentum of force of the vitality in the parts which are not cooled is expended, as before, in mechanical motion; but the whole action of the inspired oxygen is exerted on the cooled part.

If we imagine an iron cylinder, into which we admit steam under a certain pressure, then if the force with which the particles of the iron cohere be equal to the force which tends to separate them, an equilibrium will result ; that is, the whole effect of the steam will be neutralized by the resistance. But if one of the sides of the cylinder be moveable, a piston-rod, for example, and offer to the pressure of the steam a less resistance than other parts, the whole force will be expended in moving this one side—that is, in raising the piston-rod. If we do not introduce fresh steam (fresh force), an equilibrium will soon be established. The piston-rod resists a certain force without moving, but is raised by an increased pressure. When this excess of force has been consumed in motion, it cannot be raised higher ; but if new vapour be continually admitted, the rod will continue to move.

In the cooled part of the body, the living tissues offer a less resistance to the chemical action of the inspired oxygen ; the power of the oxygen to unite with the elements of the tissues is, at this part, exalted. When the part has once lost its condition of life, resistance entirely ceases ; and in consequence of

the combination of the oxygen with the elements of
the metamorphosed tissues, a greater amount of
heat is liberated.

For a given amount of oxygen, the heat produced
is, in all cases, exactly the same. In the cooled
part, the change of matter, and with it the disengagement of heat, increases; while in the other
parts the change of matter and liberation of heat
decrease. But when the cooled part, by the union
of oxygen with the elements of the metamorphosed
tissues, has recovered its original temperature, the
resistance of its living particles to the oxygen conveyed to them again increases, and, as the resistance
of other parts is now diminished, a more rapid
change of matter now occurs in them, their temperature rises, and along with this, if the cause of the
change of matter continue to operate, a larger
amount of vital force becomes available for mechanical purposes.

Let us now suppose that heat is abstracted from
the whole surface of the body; in this case the
whole action of the oxygen will be directed to the
skin, and in a short time the change of matter
must increase throughout the body. Fat, and·all
such matters as are capable of combining with the
oxygen which is brought to them in larger quantity
than usual, will be expelled from the body in the
form of oxidised compounds.

III.

THEORY OF DISEASE.

Every substance or matter, every chemical or mechanical agency, which changes or disturbs the restoration of the equilibrium between the manifestations of the causes of waste and supply, in such a way as to add its action to the causes of waste, is called *a cause of disease*. *Disease* occurs when the sum of vital force, which tends to neutralize all causes of disturbance (in other words, when the resistance offered by the vital force), is weaker than the acting cause of disturbance.

Death is that condition in which all resistance on the part of the vital force entirely ceases. So long as this condition is not established, the living tissues continue to offer resistance.

To the observer, the action of a cause of disease exhibits itself in the disturbance of the proportion between waste and supply which is proper to each period of life. In medicine, every abnormal condition of supply or of waste, in all parts or in a single part of the body, is called disease.

It is evident that one and the same cause of disease will produce in the organism very different effects, according to the period of life ; and that a certain amount of disturbance, which produces disease in the adult state, may be without influence in childhood or in old age. A cause of disease may, when it is added to the cause of waste in old age, produce death (annihilate all resistance on the part of the vital force) ; while in the adult state it may produce only a disproportion between supply and waste ; and in infancy, only an equilibrium between supply and waste (the abstract state of health).

A cause of disease which strengthens the causes of supply, either directly, or indirectly by weakening the action of the causes of waste, destroys, in the child and in the adult, the relative normal state of health ; while in old age it merely brings the waste and supply into equilibrium.

A child, lightly clothed, can bear cooling by a low external temperature without injury to health ; the force available for mechanical purposes and the temperature of its body increase with the change of matter which follows the cooling ; while a high temperature, which impedes the change of matter, is followed by disease.

On the other hand, we see, in hospitals and charitable institutions (in Brussels, for example) in which old people spend the last years of life, when the temperature of the dormitory, in winter, sinks 2 or 3 degrees below the usual point, that by this slight degree of cooling the death of the oldest and weakest, males as well as females, is brought about. They are found lying tranquilly in bed, without the slightest symptoms of disease, or of the usual recognizable causes of death.

A deficiency of resistance, in a living part, to the causes of waste is, obviously, a deficiency of resistance to the action of the oxygen of the atmosphere.

When, from any cause whatever, this resistance diminishes in a living part, the change of matter increases in an equal degree.

Now, since the phenomena of motion in the animal body are dependant on the change of matter, the increase of the change of matter in any part is followed by an increase of all motions. According to the conducting power of the nerves, the available force is carried away by the nerves of involuntary motion alone, or by all the nerves together.

Consequently, if, in consequence of a diseased transformation of living tissues, a greater amount of force be generated than is required for the production of the normal motions, it is seen in an acceleration of all or some of the involuntary motions, as well as in a higher temperature of the diseased part.

This condition is called *fever*.

When a great excess of force is produced by change of matter, the force, since it can only be consumed by motion, extends itself to the apparatus of voluntary motion.

This state is called *a febrile paroxysm*.

In consequence of the acceleration of the circulation in the state of fever, a greater amount of arterial blood, and, consequently, of oxygen, is conveyed to the diseased part, as well as to all other parts; and if the active force in the healthy parts continue uniform, the whole action of the excess of oxygen must be exerted on the diseased part alone.

According as a single organ, or a system of organs, is affected, the change of matter extends to one part alone, or to the whole affected system.

Should there be formed, in the diseased parts, in consequence of the change of matter, from the elements of the blood or of the tissue, new products, which the neighbouring parts cannot employ for their own vital functions ;—should the surrounding parts, moreover, be unable to convey these products to other parts, where they may undergo transformation, then these new products will suffer, at the place where they have been formed, a process of decomposition analogous to fermentation or putrefaction.

In certain cases, medicine removes these diseased conditions, by exciting in the vicinity of the diseased part, or in any convenient situation, an artificial diseased state (as by blisters, sinapisms, or setons) ; thus diminishing, by means of artificial disturbance, the resistance offered to the external causes of change in these parts by the vital force. The physician succeeds in putting an end to the original diseased condition, when the disturbance artificially excited (or the diminution of resistance in another part) exceeds in amount the diseased state to be overcome.

The accelerated change of matter and the elevated temperature in the diseased part shew, that the resistance offered by the vital force to the

action of oxygen is feebler than in the healthy
state. But this resistance only ceases entirely when
death takes place. By the artificial diminution of
resistance in another part, the resistance in the diseased organ is not indeed directly strengthened; but
the chemical action (the cause of the change of
matter) is diminished in the diseased part, being directed to another part, where the physician has succeeded in producing a still more feeble resistance
to the change of matter (to the action of oxygen).

A complete cure of the original disease occurs,
when external action and resistance, in the diseased
part, are brought into equilibrium. Health and the
restoration of the diseased tissue to its original condition follow, when we are able so far to weaken
the disturbing action of oxygen, by any means, that
it becomes inferior to the resistance offered by the
vital force, which, although enfeebled, has never
ceased to act; for this proportion between these
causes of change is the uniform and necessary condition of increase of mass in the living organism.

In cases of a different kind, where artificial external disturbance produces no effect, the physician
adopts other indirect methods to exalt the resistance offered by the vital force. These methods, the
result of ages of experience, are such, that the most
perfect theory could hardly have pointed them out
more acutely or more justly than has been done
by the observation of sagacious practitioners. He
diminishes, by blood-letting, the number of the

carriers of oxygen (the globules), and by this means the conditions of change of matter; he excludes from the food all such matters as are capable of conversion into blood; he gives chiefly or entirely non-azotised food, which supports the respiratory process, as well as fruit and vegetables, which contain the alkalies necessary for the secretions.

If he succeed, by these means, in diminishing the action of the oxygen in the blood on the diseased part, so far that the vital force of the latter, its resistance, in the smallest degree overcomes the chemical action; and if he accomplish this, without arresting the functions of the other organs, then restoration to health is certain.

To the method of cure adopted in such cases, if employed with sagacity and acute observation, there is added, as we may call it, an ally on the side of the diseased organ, and this is the vital force of the healthy parts. For, when blood is abstracted, the external causes of change are diminished also in them, and their vital force, formerly neutralized by these causes, now obtains the preponderance. The change of matter, indeed, is diminished throughout the body, and with it the phenomena of motion; but the sum of all resisting powers, taken together, increases in proportion as the amount of the oxygen acting on them in the blood is diminished. In the sensation of *hunger*, this resistance, in a certain sense, makes itself known; and the preponderating vital force exhibits itself, in many patients

s 2

when hunger is felt, in the form of an abnormal growth, or an abnormal metamorphosis of certain parts of organs. *Sympathy* is the transference of diminished resistance from one part, not exactly to the next, but to more distant organs, when the functions of both mutually influence each other. When the action of the diseased organ is connected with that of another—when, for example, the one no longer produces the matters necessary to the performance of the functions of the other—then the diseased condition is transferred, but only apparently, to the latter.

In regard to the nature and essence of the vital force, we can hardly deceive ourselves, when we reflect, that it behaves, in all its manifestations, exactly like other natural forces ; that it is devoid of consciousness or of volition, and is subject to the action of a blister.

The nerves, which accomplish the voluntary and involuntary motions in the body, are, according to the preceding exposition, not the producers, but only the conductors of the vital force ; they propagate motion, and behave towards other causes of motion, which in their manifestations are analogous to the vital force, towards a current of electricity, for example, in a precisely analogous manner. They permit the current to traverse them, and present, as conductors of electricity, all the phenomena which they exhibit as conductors of the vital force. In the present state of our knowledge, no one, probably, will imagine that electricity is to be considered as the cause of the phenomena of motion in the body; but still, the medicinal action of electricity, as well as that of a magnet, which, when placed in contact with the body, produces a current of electricity, cannot be denied. For to the existing force of motion or of disturbance there is added, in the electrical current, a new cause of motion and of change in form and structure, which cannot be considered as altogether inefficient.

Practical medicine, in many diseases, makes use of cold in a highly rational manner, as a means of exalting and accelerating, in an unwonted degree, the change of matter. This occurs especially in certain morbid conditions in the substance of the centre of the apparatus of motion; when a glowing heat and a rapid current of blood towards the head point out an abnormal metamorphosis of the brain. When this condition continues beyond a certain time, experience teaches that all motions in the body cease. If the change of matter be chiefly confined to the brain, then the change of matter, the generation of force, diminishes in all other parts. By surrounding the head with ice, the temperature is lowered, but the cause of the liberation of heat continues; the metamorphosis, which decides the issue of the disease, is limited to a short period. We must not forget, that the ice melts and absorbs heat from the diseased part; that if the ice be removed before the completion of the metamorphosis, the temperature

again rises; that far more heat is removed by means
of ice than if we were to surround the head with a
bad conductor of heat. There has obviously been
liberated in an equal time a far larger amount of
heat than in the state of health; and this is only
rendered possible by an increased supply of oxygen,
which must have determined a more rapid change
of matter.

The self-regulating steam-engines, in which, to
produce a uniform motion, the human intellect has
shewn the most admirable acuteness and sagacity,
furnish no unapt image of what occurs in the animal
body.

Every one knows, that in the tube which conveys
the steam to the cylinder where the piston-rod is
to be raised, a stop-cock of peculiar construction
is placed, through which all the steam must pass.
By an arrangement connected with the regulating
wheel, this stop-cock opens when the wheel moves
slower, and closes more or less completely when
the wheel moves faster than is required for a
uniform motion. When it opens, more steam is
admitted (more force), and the motion of the machine is accelerated. When it shuts, the steam is
more or less cut off, the force acting on the piston-
rod diminishes, the tension of the steam increases,
and this tension is accumulated for subsequent use.
The tension of the vapour, or the force, so to speak,
is produced by change of matter, by the combustion
of coals in the fire-place. The force increases (the

amount of steam generated and its tension increase)
with the temperature in the fire-place, which depends on the supply of coals and of air. There are
in these engines other arrangements, all intended
for regulation. When the tension of steam in the
boiler rises beyond a certain point, the passages for
admission of air close themselves ; the combustion
is retarded, the supply of force (of steam) is diminished. When the engine goes slower, more steam
is admitted to the cylinder, its tension diminishes,
the air passages are opened, and the cause of disengagement of heat (or production of force) increases. Another arrangement supplies the fire-
place incessantly with coals in proportion as they
are wanted.

If we now lower the temperature at any part of
the boiler, the tension within is diminished ; this is
immediately seen in the regulators of force, which
act precisely as if we had removed from the boiler
a certain quantity of steam (force). The regulator
and the air passages open, and the machine supplies
itself with more coals.

The body, in regard to the production of heat and
of force, acts just like one of these machines. With
the lowering of the external temperature, the respirations become deeper and more frequent ; oxygen
is supplied in greater quantity and of greater density ; the change of matter is increased, and more
food must be supplied, if the temperature of the
body is to remain unchanged.

It is hardly necessary to mention, that in the body, the tension of vapour cannot, any more than an electrical current, be considered the cause of the production of force.

From the theory of disease developed in the preceding pages, it follows obviously, that a diseased condition once established, in any part of the body, cannot be made to disappear by the chemical action of a remedy. A limit may be put by a remedy to an abnormal process of transformation ; that process may be accelerated or retarded ; but this alone does not restore the normal (healthy) condition.

The art of the physician consists in the knowledge of the means which enable him to exercise an influence on the *duration* of the disease ; and in the removal of all disturbing causes, the action of which strengthens or increases that of the actual cause of disease.

It is only by a just application of its principles that any theory can produce really beneficial results. The very same method of cure may restore health in one individual, which, if applied to another, may prove fatal in its effects. Thus in certain inflammatory diseases, and in highly muscular subjects, the antiphlogistic treatment has a very high value; while in other cases blood-letting produces unfavourable results. The vivifying agency of the blood must ever continue to be the most important condition in the restoration of a disturbed equilibrium, which result is always dependant on the saving of time; and the blood must, therefore, be considered and constantly kept in view, as the ultimate and most powerful cause of a lasting vital resistance, as well in the diseased as in the unaffected parts of the body.

It is obvious, moreover, that in all diseases where the formation of contagious matter and of exanthemata is accompanied by fever, two diseased conditions simultaneously exist, and two processes are simultaneously completed; and that the blood, as it were by re-action (*i. e.* fever), becomes a means of cure, as being the carrier of that substance (oxygen) without the aid of which the diseased products cannot be rendered harmless, destroyed, or expelled from the body; a means of cure by which, in short, neutralization or equilibrium is effected.

IV.

THEORY OF RESPIRATION.

During the passage of the venous blood through the lungs, the globules change their colour; and with this change of colour, oxygen is absorbed from the atmosphere. Further, for every volume of oxygen absorbed, an equal volume of carbonic acid is, in most cases, given out.

The red globules contain a *compound of iron;* and no other constituent of the body contains iron.

Whatever change the other constituents of the

blood undergo in the lungs, thus much is certain, that the globules of venous blood experience a change of colour, and that this change depends on the action of oxygen.

Now we observe that the globules of arterial blood retain their colour in the larger vessels, and lose it only during their passage through the capillaries. All those constituents of venous blood, which are capable of combining with oxygen, take up a corresponding quantity of it in the lungs. Experiments made with arterial serum have shewn, that when in contact with oxygen it does not diminish the volume of that gas. Venous blood, in contact with oxygen, is reddened, while oxygen is absorbed; and a corresponding quantity of carbonic acid is formed.

It is evident that the change of colour in the venous globules depends on the combination of some one of their elements with oxygen; and that this absorption of oxygen is attended with the separation of a certain quantity of carbonic acid gas.

This carbonic acid is not separated from the serum; for the serum does not possess the property, when in contact with oxygen, of giving off carbonic acid. On the contrary, when separated from the globules, it absorbs from half its volume to an equal volume of carbonic acid, and, at ordinary temperatures, is not saturated with that gas. (See the article " Blut " in the " Handwörterbuch der Chemie von Poggendorff, Wöhler, and Liebig, p. 877.)

Arterial blood, when drawn from the body, is soon altered; its florid colour becomes dark red. The florid blood, which owes its colour to the globules, becomes dark by the action of carbonic acid, and this change of colour affects the globules, for florid blood absorbs a number of gases which do not dissolve in the fluid part of the blood when separated from the globules. *It is evident, therefore, that the globules have the power of combining with gases.*

The globules of the blood change their colour in different gases; and this change may be owing either to a combination or to a decomposition.

Sulphuretted hydrogen turns them blackish green and finally black; and the original red colour cannot, in this case, be restored by contact with oxygen. Here a decomposition has obviously taken place.

The globules darkened by carbonic acid become again florid in oxygen, with disengagement of carbonic acid. The same thing takes place in nitrous oxide. It is clear that they have here undergone no decomposition, and, consequently, they possess the power of combining with gases, *while the compound they form with carbonic acid is destroyed by oxygen.* When left to themselves, out of the body, the compound formed with oxygen again becomes dark, but does not recover its florid colour a second time by the action of oxygen.

The globules of the blood contain a compound of iron. From the never-failing presence of iron in red blood, we must conclude, that it is unquestionably necessary to animal life; and, since physiology has proved, that the globules take no share in the process of nutrition, it cannot be doubted that they play a part in the process of respiration.

The compound of iron in the globules has the characters of an oxidised compound; for it is decomposed by sulphuretted hydrogen, exactly in the same way as the oxides or other analogous compounds of iron. By means of diluted mineral acids, peroxide (sesquioxide) of iron may be extracted, at the ordinary temperature, from the fresh or dried red colouring matter of the blood.

The characters of the compounds of iron may, perhaps, assist us to explain the share which that metal takes in the respiratory process. No other metal can be compared with iron, for the remarkable properties of its compounds.

The compounds of protoxide of iron possess the property of depriving other oxidised compounds of oxygen; while the compounds of peroxide of iron, under other circumstances, give up oxygen with the utmost facility.

Hydrated peroxide of iron, in contact with organic matters destitute of sulphur, is converted into carbonate of the protoxide.

Carbonate of protoxide of iron, in contact with water and oxygen, is decomposed; all the carbonic acid is given off, and, by absorption of oxygen, it passes into the hydrated peroxide, which may again be converted into a compound of the protoxide.

Not only the oxides of iron, but also the cyanides of that metal, exhibit similar properties. Prussian blue contains iron in combination with all the organic elements of the body; hydrogen and oxygen (water), carbon and nitrogen (cyanogen).

When it is exposed to light, cyanogen is given off, and it becomes white; in the dark it attracts oxygen, and recovers its blue colour.

All these observations, taken together, lead to the opinion that the globules of arterial blood contain a compound of iron saturated with oxygen, which, in the living blood, loses its oxygen during its passage through the capillaries. The same thing occurs when it is separated from the body, and begins to undergo decomposition (to putrefy). The compound, rich in oxygen, passes, therefore, by the loss of oxygen (reduction), into one far less charged with that element. One of the products of oxidation formed in this process is carbonic acid. The compound of iron in the venous blood possesses the property of combining with carbonic acid; and it is obvious, that the globules of the arterial blood, after losing a part of their oxygen, will, if they meet with carbonic acid, combine with that substance.

When they reach the lungs, they will again take up the oxygen they have lost; for every volume of oxygen absorbed, a corresponding volume of carbonic acid will be separated; they will return to their former state; that is, they will again acquire the power of giving off oxygen.

For every volume of oxygen which the globules
can give off, there will be formed (as carbonic acid
contains its own volume of oxygen, without conden-
sation) neither more nor less than an equal volume
of carbonic acid. For every volume of oxygen
which the globules are capable of absorbing, no
more carbonic acid can possibly be separated than
that volume of oxygen can produce.

When carbonate of protoxide of iron, by the
absorption of oxygen, passes into the hydrated
peroxide, there are given off, for every volume of
oxygen necessary to the change from protoxide to
peroxide, four volumes of carbonic acid gas.

But from one volume of oxygen only one volume
of carbonic acid can be *produced ;* and the absorption
of one volume of oxygen can only cause, directly,
the *separation* of an equal volume of carbonic acid.
Consequently, the substance or compound which has
lost its oxygen, during the passage of arterial into
venous blood, must have been capable of absorbing or
combining with carbonic acid ; and we find, in point
of fact, that the living blood is never, in any state,
saturated with carbonic acid; that it is capable of
taking up an additional quantity, without any apparent disturbance of the function of the globules.
Thus, for example, after drinking effervescing wines,
beer, or mineral waters, more carbonic acid must
necessarily be expired than at other times. In all
cases, where the oxygen of the arterial globules has
been partly expended, otherwise than in the forma-

tion of carbonic acid, the amount of this latter gas expired will correspond exactly with that which has been formed; less, however, will be given out after the use of fat and of still wines, than after champagne.

According to the views now developed, the globules of arterial blood, in their passage through the capillaries, yield oxygen to certain constituents of the body. A small portion of this oxygen serves to produce the change of matter, and determines the separation of living parts and their conversion into lifeless compounds, as well as the formation of the secretions and excretions. The greater part, however, of the oxygen is employed in converting into oxidised compounds the newly formed substances, which no longer form part of the living tissues.

In their return towards the heart, the globules which have lost their oxygen combine with carbonic acid, producing venous blood; and, when they reach the lungs, an exchange takes place between this carbonic acid and the oxygen of the atmosphere.

The organic compound of iron, which exists in venous blood, recovers in the lungs the oxygen it has lost, and, in consequence of this absorption of oxygen, the carbonic acid in combination with it is separated.

All the compounds present in venous blood, which have an attraction for oxygen, are converted, in the lungs, like the globules, into more highly oxidised compounds; a certain amount of carbonic acid is formed, of which a part always remains dissolved in the serum of the blood.

The quantity of carbonic acid dissolved, or of that combined with soda, must be equal in venous and arterial blood, since both have the same temperature; but arterial blood, when drawn, must, after a short time, contain a larger quantity of carbonic acid than venous blood, because the oxygen of the globules is expended in producing that compound.

Hence, in the animal organism, two processes of oxidation are going on; one in the lungs, the other in the capillaries. By means of the former, in spite of the degree of cooling, and of the increased evaporation which takes place there, the constant temperature of the lungs is kept up; while the heat of the rest of the body is supplied by the latter.

A man, who expires daily 13·9 oz. of carbon, in the form of carbonic acid, consumes, in 24 hours, 37 oz. of oxygen, which occupy a space equal to 807 litres = 51,648 cubic inches (hessian).

If we reckon 18 respirations to a minute, we have, in 24 hours, 25,920 respirations; and, consequently, in each respiration, there are taken into the blood $\frac{51648}{25920}$ = 1·99 cubic inch of oxygen.

In one minute, therefore, there are added to the constituents of the blood $18 \times 1·99 = 35·8$ cubic inches of oxygen, which, at the ordinary temperature, weigh rather less than 12 grains.

If we now assume, that in one minute 10 lbs. of blood pass through the lungs (Müller, Physiologie,

vol. i. p. 345), and that this quantity of blood measures 320 cubic inches, then 1 cubic inch of oxygen
unites with 9 cubic inches of blood, very nearly.

According to the researches of Dénis, Richardson,
and Nasse (Handwörterbuch der Physiologie, vol. i.
p. 138), 10,000 parts of blood contain 8 parts of peroxide of iron.   Consequently, 76,800 grains (10 lbs.
hessian) of blood contain 61·54 grains of peroxide
of iron in arterial blood, = 55·14 of protoxide in
venous blood.

Let us now assume that the iron of the globules
of venous blood is in the state of protoxide.   It
follows, that 55·14 grains of protoxide of iron, in
passing through the lungs, take up, in one minute,
6·40 grains of oxygen (the quantity necessary to
convert it into peroxide).   But since, in the same
time, the 10 lbs. of blood have taken up 12 grains
of oxygen, there remain 5·60 grains of oxygen,
which combine with the other constituents of the
blood.

Now, 55·14 grains of protoxide of iron combine
with 34·8 grains of carbonic acid, which occupy the
volume of 73 cubic inches.   It is obvious, therefore,
that the amount of iron present in the blood, if in
the state of protoxide, is sufficient to furnish the
means of carrying or transporting twice as much
carbonic acid as can possibly be formed by the
oxygen absorbed in the lungs.

The hypothesis just developed rests on well-known
observations, and, indeed, explains completely the

T

process of respiration, as far as it depends on the globules of the blood. It does not exclude the opinion that carbonic acid may reach the lungs in other ways; that certain other constituents of the blood may give rise to the formation of carbonic acid in the lungs. But all this has no connection with that vital process by which the heat necessary for the support of life is generated in every part of the body. Now it is this alone which, for the present, can be considered as the object truly worthy of investigation. It is not, indeed, uninteresting to inquire, why dark blood becomes florid by the action of nitre, common salt, &c.; but this question has no relation to the natural respiratory process.

The frightful effects of sulphuretted hydrogen, and of prussic acid, which, when inspired, put a stop to all the phenomena of motion in a few seconds, are explained in a natural manner by the well-known action of these compounds on those of iron, when alkalies are present; and free alkali is never absent in the blood.

Let us suppose that the globules lose their property of absorbing oxygen, and of afterwards giving up this oxygen and carrying off the resulting carbonic acid; such a hypothetical state of disease must instantly become perceptible in the temperature and other vital phenomena of the body. The change of matter will be arrested, while yet the vital motions will not be instantly stopped.

The conductors of force, the nerves, will convey,

as before, to the heart and intestines the power necessary for their functions. This power they will receive from the muscular system, while, as no change of matter takes place in the latter, the supply must soon fail. As no change of matter occurs, no lifeless compounds are separated, neither bile nor urine can be formed; and the temperature of the body must sink.

This state of matters soon puts a stop to the process of nutrition, and, sooner or later, death must follow, but unaccompanied by febrile symptoms, which, in this case, is a very important fact.

This example has been selected in order to shew the importance and probable advantage of an examination of the blood in analogous diseased conditions. It cannot be, in the slightest degree, doubtful that the function ascribed to the blood globules may be considered as fully explained and cleared up, if, in such morbid conditions, we shall discover a change in their form, structure, or chemical characters, a change which must be recognizable by the use of appropriate re-agents.

If we consider the force which determines the vital phenomena as a property of certain substances, this view leads of itself to a new and more rigorous consideration of certain singular phenomena, which these very substances exhibit, in circumstances in which they no longer make a part of living organisms.

APPENDIX;

CONTAINING

THE ANALYTICAL EVIDENCE

REFERRED TO IN THE SECTIONS IN WHICH ARE DESCRIBED

THE

CHEMICAL PROCESSES OF RESPIRATION, OF NUTRITION,

AND OF THE

METAMORPHOSIS OF TISSUES.

---

*_** *The Notes correspond with the numbers in parentheses in the text. All the Analyses quoted, which have the mark * attached, have been made in the chemical laboratory of the University of Giessen.*

APPENDIX.

INTRODUCTION TO THE ANALYSES.

THE method formerly employed to exhibit the differences in composition of different substances, that, namely, of giving the proportions of the various elements in 100 parts, has been long abandoned by chemists; because it affords no insight into the relations which exist between two or more compounds. In order to give some proofs of this statement, we shall here state, in that form, the composition of aldehyde and acetic acid, of oil of bitter almonds and benzoic acid.

	Acetic acid.	Aldehyde.	Benzoic acid.	Oil of bitter almonds.
Carbon	40·00	55·024	69·25	79·56
Hydrogen ...	6·67	8·983	4·86	5·56
Oxygen	53·33	35·993	25·89	14·88

Now aldehyde is converted into acetic acid, and oil of bitter almonds into benzoic acid, simply by the addition of oxygen, without any change in regard to the other elements. This important relation cannot be traced in the mere numerical results of analysis as above given; but if the composition of the related compounds be expressed in formulæ, according to equivalents, the connection in each case becomes obvious, even to him who knows no more of chemistry than that C represents an equivalent or combining proportion of carbon, H an equivalent of hydrogen, and O an equivalent of oxygen.

Formula		Formula	
of acetic acid.	of aldehyde.	of benzoic acid.	of oil of bitter almonds.
$C_4H_4O_4$.	$C_4H_4O_2$.	$C_{14}H_6O_4$.	$C_{14}H_6O_2$.

These formulæ are exact expressions of the results of analysis, which, in each of the two cases quoted, refer to a fixed quantity of carbon; in one to 4 equivalents, in the other to 14. They shew, that acetic acid differs from aldehyde, and benzoic acid from oil of bitter almonds, only in the proportion of oxygen.

Nor is it more difficult to understand the signification of the following formulæ.

$$\text{Cyamelide.} \qquad \text{1 eq. cyanuric acid.} \qquad \text{3 eq. hydrated cyanic acid.}$$
$$C_6N_3H_3O_6 = Cy_3(=C_6N_3)O_3 + 3HO = 3(CyO + HO) =$$
$$= C_6N_3H_3O_6 \qquad\qquad = C_6N_3H_3O_6.$$

(In these formulæ, N represents an equivalent of nitrogen, and Cy an equivalent of cyanogen. This latter substance being composed of 2 equivalents of carbon and 1 eq. of nitrogen, $Cy = C_2N$.)

The first formula (that of cyamelide) is what is called an empirical formula, in which the relative proportions of the elements are, indeed, exactly known, but where we have not even a theory, far less any actual knowledge, of the order in which they are arranged. The second formula is intended to express the opinion that 3 eq. of cyanogen (= 6 eq. of carbon + 3 eq. of nitrogen) having united to form a compound atom or molecule, have combined with 3 eq. of oxygen and 3 eq. of water, to form 1 eq. of hydrated cyanuric acid. The third expresses the order in which the elements are supposed to be arranged in hydrated cyanic acid, the whole multiplied by 3. Each equivalent of cyanic acid is formed of 1 eq. of cyanogen, 1 eq. of oxygen, and 1 eq. of water; and hence the same number of atoms of each element, which together formed 1 eq. of cyanuric acid, is here so divided as to yield 3 eq. of cyanic acid.

We have here, therefore, the same absolute and relative amount of atoms of each element, arranged in three different ways; yet in each of these the proportions of the elements, calculated for 100 parts, must of course be the same. It is easy, therefore, to see the advantage we possess by the use of formulæ; that, namely, of exhibiting the relations existing between compounds of different composition; and that also of expressing the actual, probable, or possible differences between substances whose composition, in 100 parts, is the same, while their properties, as in the case above quoted, are perfectly distinct.

It does not come within our province here to explain the method or rule by which the composition of a substance, in 100 parts (as it is always obtained in analysis), is expressed in a formula; we shall only describe the rule for calculating, from a given formula, the composition in 100 parts. For this purpose it must be noted that C, in a chemical formula, signifies a weight of carbon expressed by the number 76·437 (according to the most recent determinations 75·8 or 75·0, a variation which has no effect whatever on the formulæ here adduced, all of which are calculated on the number 76·437); that H signifies a weight of hydrogen = 12·478; N a weight of nitrogen = 177·04; and lastly O a weight of oxygen = 100.

The formula of proteine, $C_{48}N_6H_{36}O_{14}$, expresses, therefore,

48 times 76·437 = 3668·88 carbon,
6 times 177.040 = 1062·24 nitrogen,
36 times 12·478 = 449·26 hydrogen,
14 times 100·000 = 1400·00 oxygen.

The sum gives a weight of 6580·38 proteine.

Therefore—

			In 100 parts.
In 6580·38 parts of proteine are contained	3668·88 carbon		55·742
In 6580·38	ditto	1062·24 nitrogen	16·143
In 6580·38	ditto	449·26 hydrogen	6·827
In 6580·38	ditto	1400·00 oxygen	21·288
			100·000

The actual results of analysis, reduced to 100 parts, when compared with the above numbers, will shew how far the assumed formula is correct; or, supposing the formula ascertained, they will shew the degree of accuracy displayed by the experimenter. Thus the proportions in 100 parts, calculated from the formula, furnish an important check to the operator, and, conversely, the formula calculated from his results, when compared with other known formulæ, supplies a test of his accuracy, or of the purity of the substance analyzed.

NOTE (1), p. 12.

CONSUMPTION OF OXYGEN BY AN ADULT.

An adult man

According to	consumes of oxygen in 24 hours		produces of carbonic acid in 24 hours		Carbon contained in the carbonic acid.
	cubic in.	grains.	cubic in.	grains.	grains.
Lavoisier and Seguin	46,037	15,661	14,930	8,584	2,820 French.
Menzies	51,480	17,625			English.
Davy	45,504	15,751	31,680	17,811	4,853 do.
Allen and Pepys ...	39,600	13,464	39,600	18,612	5,148 do.

NOTE (2), p. 13.

COMPOSITION OF DRY BLOOD (see note 28).

	In 100 parts.	In 4·8 lbs. Hessian = 36,864 grains.
Carbon......	51·96	19154·5
Hydrogen...	7·25	2672·7
Nitrogen ...	15·07	5555·4
Oxygen ...	21·30	7852·0
Ashes	4·42	1629·4
	100·00	36864·0

Grains.   Grains.

19154·5 carbon form, with 50539·5 oxygen, carbonic acid.

2672·7 hydrogen do.   21415·8   do.   water.

Sum = 71955·3   do.

Deduct oxygen present in blood .... } = 7852·0

Remain ... 64103·3 grains of oxygen, required for the complete combustion of 4·8 lbs. of dry blood.

It is assumed, in this calculation, that 24 lbs. of blood yield 4·8 lbs. (20 per cent.) of dry residue. The remainder, 80 per cent., is water.

NOTE (3), p. 14.

DETERMINATION OF THE AMOUNT OF CARBON EXPIRED.

1. ANALYSIS OF

Fæces.

2·356 dry fæces left 0·320 ashes (13·58 per cent.)
0·352 dry fæces yielded 0·576 carbonic acid, and 0·218 water.

Lentils.

0·566 lentils, dried at 212°, yielded 0·910 carbonic acid, and 0·366 water.

Pease.

1·060 pease, dried at 212°, left 0·037 ashes.
0·416 do.   do.   yielded 0·642 carbonic acid, and 0·241 water.

Potatoes.

0·443 dried potatoes yielded 0·704 carbonic acid, and 0·248 water.

Black Bread (Schwarzbrod).

0·302 dried black bread yielded 0·496 carbonic acid, and 0·175 water.
0·241   do.   0·393   do.   0·142 water.

From the above, which are the direct results of experiment, the composition in 100 parts is calculated as in the following table.

2. Composition

	Of Fæces.	Of Black Bread.		Of Potatoes.		Of Flesh.
	Playfair.*	Bœckmann.*		Boussingault. Bœckmann.*		
Carbon ...	45·24	45·09	45·41	44·1	43·944	(See note
Hydrogen	6·88	6·54	6·45	5·8	6·222	28.)
Nitrogen \\ Oxygen /	34·73	45·12	44·89	45·1	44·919	
Ashes ...	13·15	3·25	3·25	5·0	4·915	
	100·00	100·00	100·00	100·0	100 000	
Water ...	300·00					
	400·00					

	Of Pease.	Of Lentils.	Of Beans.
	Playfair.*	Playfair.*	Playfair.*
Carbon...............	35·743	37·38	38·24
Hydrogen	5·401	5·54	5·84
Nitrogen \\ Oxygen /	39·366	37·98	38·10
Ashes	3·490	3·20	3·71
Water	16·000	15·90	14·11
	100·000	100·00	100·00

	Fresh Meat.		Potatoes.		Black Bread.	
	Bœckmann.*		Boussingault.		Bœckmann.*	
Water.........	75	74·8	72·2	73·2	33	31·418
Dry Matter	25	25·2	27·8	26·8	67	68·592
	100	100·0	100·0	100·0	100	100·000

3. Calculation,

with the help of the preceding data, of the amount of carbon expired by an adult man. The following results are deduced from observations made (see table) on the average daily consumption of food, by from 27 to 30 soldiers in barracks for a month, or by 855 men for one day. The food, consisting of bread, potatoes, meat, lentils, pease, beans, &c., was weighed, with the utmost exactness, every day during a month (including even pepper, salt, and butter); and each article of food was separately subjected to ultimate analysis. The only exceptions, among the men, to the uniform allowance of food, were three soldiers of the guard, who, in addition to the daily allowance of 2 lbs. of bread, received, during each of the periods allotted for the pay of the troops, $2\frac{1}{2}$ lbs. extra; and one drummer, who, in the same period, left $2\frac{1}{2}$ lbs. unconsumed. According to an approximative report by the sergeant-major, each soldier consumes daily, on an average, out of barracks, 3 oz. of sausage, $\frac{3}{4}$ oz. of butter, $\frac{1}{2}$ pint of beer, and $\frac{1}{16}$ pint of brandy; the carbon of which articles amounts to more than double that of the fæces and urine taken together. In the soldier, the fæces amount daily, on an average, to $5\frac{1}{2}$ oz.; they contain 75 per cent. of water, and the dry residue contains 45·24 per cent. of carbon, and 13·15 per cent. of ashes. 100 parts of fresh fæces consequently contain 11·31 per cent. of carbon, very nearly the same proportion as in fresh meat. In the calculation, the carbon of the fæces and of the urine has been assumed as equal to that of the green vegetables, and of the food (sausages, butter, beer, &c.) consumed in the alehouse.

From the observations, as recorded in the table, the following conclusions are deduced.

Flesh.—Meat devoid of fat, if reckoned at 74 per cent. water, and 26 per cent. dry matter, contains in 100 parts very nearly 13·6 parts of carbon. Ordinary meat contains both fat and cellular tissue, which together amount to $\frac{1}{4}$th of the weight of the meat as bought from the butcher. The number of ounces consumed (by 855 men) was 4,448, consisting, therefore, of

3812·5 oz. of flesh, free from fat, containing of carbon 518·5 oz.

  635·5 oz. of fat and cellular tissue,   ditto   449·0 oz.

4448·0 oz.   In all, carbon 967·5 oz.

With the bones, the meat, as purchased, contains 29 per cent. of fixed matter, including bones; 4,448 oz. of flesh therefore contain 448 oz. of dry bones. These have not been included in the calculation, although, when boiled, they yield from 8 to 10 per cent. of gelatine, which is taken as food in the soup.

Fat.—The amount of fat consumed was 56 oz.; which, the carbon being calculated at 80 per cent., contain in all 44·8 oz. of carbon.

Lentils, pease, and beans.—There were consumed 53·5 oz. of lentils, 185·5 oz. of pease, and 218 oz. of beans. Assuming the average amount of carbon in these vegetables to be 37 per cent., the total quantity of carbon consumed in this form was 169·1 oz.

Potatoes.—100 parts of fresh potatoes contain 12·2 parts of carbon. In the 15,876 oz. of potatoes consumed, therefore, the amount of carbon was 1936·85 oz.

Bread.—855 men eat daily 855 times 32 oz., besides 36 lbs. of bread in the soup, which in all amounts to 27,936 oz. 100 oz. of fresh bread contain, on an average, 30·15 oz. of carbon; consequently, the carbon consumed in the bread amounts to 8771·5 oz.

The total consumption, therefore, was,

In the meat	967·50	oz. of carbon.
In the fat	44·80	ditto
In the lentils, pease, and beans ...	169·10	ditto
In the potatoes	1936·85	ditto
In the bread	8771·50	ditto
Consumed by 855 men	11889·75	ditto
Consumed by 1 man	13·9	ditto

The fæces of a soldier weigh 5·5 oz., and contain, in the fresh state, 11 per cent. of carbon. For 86 kreutzer (about 2s. 5d. sterling) there may be bought, on an average, 172 lbs. of vegetables, such as cabbages, greens, turnips, &c.: 25 maas of sour krout weigh 100 lbs.; and for 48½ kreutzer (1s. 5d. sterling) there are bought, on an average, 24¼ lbs. of onions, leeks, celery, &c.* 855 men consumed

Of green vegetables	2,802 oz.
Of sour krout	1,600
Of onions, &c.	388
In all	4,790
And one man....................	5·6 oz.

For this reason, the carbon of the last-mentioned articles of food has been assumed as equal to that of the fæces and urine. Sausages, brandy, beer, in short, the small quantity of food taken irregularly in the alehouse, has not been included in the calculation.

The daily allowance of bread, being uniformly 2 lbs. per man, with the exceptions formerly mentioned, has not been inserted in the table, which includes only those matters of which, from the daily allowance being variable, an average was required. The small quantity of bread in the table is that given in the soup, which is over and above the daily supply.

* In the original table, the quantities of these vegetables are entered according to their value in kreutzers, but they are here calculated by weight from the above data, as this appeared better adapted for comparison in this country than the prices would have been.—Ed.

TABLE I. (to Note 3).

Containing a Summary of the Victuals consumed during November, 1840, by a Company of the Body Guard of the Grand Duke of Hesse Darmstadt.

1840. November, in the period from the		Number of men supplied with food.	Beef.	Pork.	Potatoes.	Peas.	Beans.	Lentils.	Sourkrout.	Green Vegetables.	Bread in Soup.	Salt.	Onions, Leeks, &c.	Pepper. Price in Kreutzers. 3kr.=1d.	Fat or Lard.	Vinegar.
			lbs.	lbs.	lbs. oz.	lbs. oz.	lbs. oz.	lbs. oz.	lbs.	lbs.	lbs.	lbs.	lbs.	kr.	oz.	pints.
1st to 5th		139	36	9	147 0	4 15½	3 7½	...	20	12	5	4½	4	2½	13⅓	...
6th to 10th		145	37	9	165 6	...	...	...	16	70	7½	5	3¼	2½	10⅔	..
11th to 15th		136	36	9	153 2	...	3 7½	...	16	42	7½	4½	3½	2	8	...
16th to 20th		136	37	9	177 10	3 5	3 7½	...	16	12	6	4½	4¼	3½	13⅓	...
				Pork Sausages.												
21st to 25th		147	39	7½	171 8	...	...	3 5½	...	36	7½	5½	5½	2½	5⅓	1½
26th to 30th		152	30	19½	177 10	3 5	3 7½	...	32	...	2½	4	3¾	2½	5⅓	...
Total ...		855	215	63	992 4	11 9½	13 14	3 5½	100	172	36	28	20¼	15½	56	1½
The average number of men daily fed is $\frac{855}{30}$ = 28½, therefore each man had	Monthly		$7\frac{31}{37}$ lbs.	$2\frac{12}{37}$ lbs.	34lbs. 13oz.	$6\frac{87}{171}$ oz.	$7\frac{135}{171}$ oz.	$1\frac{150}{171}$ oz.	3 lbs. $11\frac{91}{151}$ oz.	$6\frac{6}{171}$ lbs.	$1\frac{15}{57}$ lb.	$\frac{56}{57}$ lb.	$11\frac{7}{19}$ oz.	$\frac{31}{57}$ kr.	$1\frac{55}{57}$ oz.	$\frac{3}{57}$ pint.
	Daily ...		$4\frac{4}{171}$ oz.	$1\frac{153}{855}$ oz.	1 lb. $2\frac{54}{95}$ oz.	$\frac{371}{1710}$ oz.	$\frac{222}{855}$ oz.	$\frac{107}{1710}$ oz.	$1\frac{149}{151}$ oz.	$3\frac{187}{855}$ oz.	$\frac{36}{855}$ lb.	$\frac{28}{855}$ lb.	$\frac{324}{855}$ oz.	$\frac{31}{1710}$ kr.	$\frac{56}{855}$ oz.	$\frac{3}{1710}$ pint.

N. B.—As all the weights mentioned in the text of this work are Hessian pounds and ounces, the different articles in this table have been reduced to the same weights, excepting the Pepper and Vinegar, the amount of both of which is so small that they may be omitted, without affecting the general result. For the benefit of those who may wish to reduce these weights to avoirdupois weight, it may here be mentioned, that 1 lb. Hessian = 16 oz. Hessian = 7680 grains Hessian = 7712 grains avoirdupois. Consequently, since the 1 lb. avoirdupois is = 7000 grains Troy, 1 lb. avoirdupois : 1 lb. Hessian :: 7000 : 7712, or as 1 : 1·1017; and 1 oz. Hessian = 482 grains Troy (1 oz. Troy is = 480 grains), while 1 oz. avoirdupois is = 437·5 grains Troy.

TABLE II.—Note (4), p. 14. *a*

FOOD CONSUMED BY A HORSE IN TWENTY-FOUR HOURS.

Articles of food.	Weight in the fresh state.	Weight in the dry state.	Carbon.	Hydrogen.	Oxygen.	Nitrogen.	Salts and earthy matters.
Hay	7500	6465	2961·0	323·2	2502·0	97·0	581·8
Oats	2270	1927	977·0	123·3	707·2	42·4	77·1
Water......	16000	...	...	...	...	...	13·3
Total ...	25770	8392	3938·0	446·5	3209·2	139·4	672·2

EXCRETIONS OF A HORSE IN TWENTY-FOUR HOURS.

Excretions.	Weight in the fresh state.	Weight in the dry state.	Carbon.	Hydrogen.	Oxygen.	Nitrogen.	Salts and earthy matters.
Urine	1330	302	108·7	11·5	34·1	37·8	109·9
Excrements	14250	3525	1364·4	179·8	1328·9	77·6	574·6
Total ...	15580	3827	1472·9	191·3	1363·0	115·4	684·5
Total from the previous part of this Table.	25770	8392	3938·0	446·5	3209·2	139·4	672·2
Difference	10190	4565	2465·1	255·2	1846·2	24·0	12·3
+ or —	—	—	—	—	—	—	+

a Boussingault, Ann. de Ch. et de Phys., LXX., 136. The weights in this table are given in grammes. 1 gramme = 15·44 grains Troy, very nearly.

TABLE II.—Note (4), p. 14 (*concluded*).

FOOD CONSUMED BY A COW IN TWENTY-FOUR HOURS.

Articles of food.	Weight in the fresh state.	Weight in the dry state.	Carbon.	Hydrogen.	Oxygen.	Nitrogen.	Salts and earthy matters.
Potatoes ...	15000	4170	1839·0	241·9	1830·6	50·0	208·5
After Grass	7500	6315	2974·4	353·6	2204·0	151·5	631·5
Water......	60000	...	...	...	...	...	50·0
Total ...	82500	10485	4813·4	595·5	4034·6	201·5	889·0

EXCRETIONS OF A COW IN TWENTY-FOUR HOURS.

Excretions.	Weight in the fresh state.	Weight in the dry state.	Carbon.	Hydrogen.	Oxygen.	Nitrogen.	Salts and earthy matters.
Excrements	28413	4000·0	1712·0	208·0	1508·0	92·0	480·0
Urine	8200	960·8	261·4	25·0	253·7	36·5	384·2
Milk	8539	1150·6	628·2	99·0	321·0	46·0	56·4
Total ...	45152	6111·4	2601·6	332·0	2082·7	174·5	920·6
Total of first part of this Table.	82500	10485·0	4813·4	595·5	4034·6	201·5	889·0
Difference	37348	4374·6	2211·8	263·5	1951·9	27·0	31·6
+ or —	—	—	—	—	—	—	+

NOTE (5), p. 19.

TEMPERATURE OF THE BLOOD AND FREQUENCY OF THE PULSE.

	The mean temperature is F.	According to Prevost and Dumas, The frequency	
		of the pulse in the minute.	of the respiration in the minute.
In the Pigeon	107·6°	136	34
Common Fowl	106·7°	140	30
Duck	108·5°	170	21
Raven	108·5°	110	21
Lark	117·2°	200	22
Simia Callitriche ...	95·9°	90	30
Guinea Pig	100·4°	140	36
Dog	99·3°	90	28
Cat	101·3°	100	24
Goat	102·5°	84	24
Hare..................	100·4°	120	36
Horse	98·2°	56	16
Man	98·6°	72	18
Man (Liebig)	97·7°	65	17
Woman (Liebig) ...	98·2°	60	15

The temperature of a child is 102·2°.

The temperature of the human body, in the mouth or in the rectum, for example, is from 97·7° to 98·6°. That of the blood (Majendie) is from 100·6° to 101·6°. As a mean temperature, 99·5° has been adopted in this work, page 19.

NOTE (6), p. 36.

The prisoners in the house of arrest at Giessen receive daily 1½ lb. of bread (24 oz.), which contain 7¼ oz. of carbon. They receive, besides, 1 lb. of soup daily, and on each alternate day, 1 lb. of potatoes.

1½ lb. of bread contains	7·25	oz. of carbon.
1 lb. of soup contains	0·75	ditto
½ lb. of potatoes contains.........	1·00	ditto
Total	9·00	ditto‡

NOTE (7), p. 43.

COMPOSITION OF THE FIBRINE AND ALBUMEN OF BLOOD. a

	Albumen from Serum of Blood. Scherer.*			Fibrine. Scherer.*		Mulder.
	I.	II.	III.	I.	II.	III.
Carbon	53·850	55·461	56·097	53·671	54·454	54·56
Hydrogen.........	6·983	7·201	6·880	6·878	7·069	6·90
Nitrogen	15·673	15·673	15·681	15·763	15·762	15·72
Oxygen ....⎫						
Sulphur......... ⎬	23·494	21·655	22·342	23·688	22·715	22·82
Phosphorus ...⎭						

a Annalen der Chem. und Pharm., XXVIII., 74, and XL., 33, 36.

For additional analyses of animal fibrine and albumen, see Note (27), which also contains analyses of the various animal tissues.

‡ At page 36 the carbon contained in the daily food of these prisoners is calculated at 8½ oz., and the appendix in the original makes the number also 8·5, apparently by an error in adding up the above numbers, which yield the sum of 9 oz. Possibly there may be an error in excess in the proportion of carbon calculated for the soup, which, in that case, ought to be 0·25 oz.—EDITOR.

NOTE (8), p. 48.

COMPOSITION OF VEGETABLE FIBRINE, VEGETABLE
ALBUMEN, VEGETABLE CASEINE, AND VEGETABLE
GLUTEN.

	VEGETABLE FIBRINE.				GLUTEN, As obtained from wheat flour.	
	Scherer.[a]			Jones.[b]	Marcet.[c]	Boussingault.
	I.	II.	III.	IV.	I.	II.
Carbon.........	53·064	54·603	54·617	53·83	55·7	53·5
Hydrogen......	7·132	7·302	7·491	7·02	14·5	15·0
Nitrogen	15·359	15·809	15·809	15·58	7·8	7·0
Oxygen ⎤ Sulphur ⎬ Phosphorus ⎦	24·445	22·285	22·083	23·56	22·0	24·5

a Ann. der Chem., und Pharm., XL., 7.
b Ibid., XL., 65.
c L. Gmelin's Theor. Chemie, II., 1092.

VEGETABLE ALBUMEN, a

	From Rye. Jones.[*]	From Wheat. Jones.[*]	From Gluten. Varrentrapp & Will.[*]	From Almonds. Jones.[*]
Carbon	54·74	55·01	54·85	57·03
Hydrogen......	7·77	7·23	6·98	7·53
Nitrogen	15·85	15·92	15·88	13·48
Oxygen ⎤ Sulphur ⎬ Phosphorus ⎦	21·64	21·84	22·39	21·96

	Boussingault.	Varrentrapp and Will.[*]
Carbon	52·7	——
Hydrogen	6·9	——
Nitrogen...............	18·4	15·70
Oxygen, &c.	22·0	——

a Ann. der Chem. und Pharm., XL., 66, and XXXIX., 291.

VEGETABLE CASEINE. a

	Scherer.*	Jones.*	Sulphate of Caseine and Potash. Varrentrapp and Will.	
Carbon	54·138	55·05	51·41	51·24
Hydrogen	7·156	7·59	7·83	6·77
Nitrogen.........	15·672	15·89	14·48	13·23
Oxygen, &c. ...	23·034	21·47	——	——

a Ann. der Chem. und Pharm., XXXIX., 291, and XL., 8 and 67.

VEGETABLE GLUTEN.

	Jones.* a	Boussingault.	
Carbon	55·22	54·2	52·3
Hydrogen	7·42	7·5	6·5
Nitrogen	15·98	13·9	18·9
Oxygen, &c.	21·38	24·4	22·3

a Ann. der Chem. und Pharm., XL., 66.

The pure gluten, analyzed by Jones, was that portion of the raw gluten from wheat flour which is soluble in hot alcohol. The insoluble portion is vegetable fibrine, the analysis of which has been already given.

NOTE (9), p. 51.

COMPOSITION OF ANIMAL CASEINE. a

	Scherer.				
	From fresh milk.	From sour milk.		From milk by acetic acid.	Albuminous substance in milk.b
	I.	II.	III.	IV.	V.
Carbon......	54·825	54·721	54·665	54·580	54·507
Hydrogen	7·153	7·239	7·465	7·352	6·913
Nitrogen ...	15·628	15·724	15·724	15·696	15·670
Oxygen. Sulphur } ...	22·394	22·316	22·146	22·372	22·910

a Ann. der Chem. und Pharm., XL., 40 et seq.

b This substance, called, in German, *zieger*, is contained in the whey of milk after coagulation by an acid. It is coagulated by heat, and very much resembles albumen.

Mulder. *a*

Carbon	54·96
Hydrogen	7·15
Nitrogen	15·80
Oxygen	21·73
Sulphur	0·36

a For the analysis of vegetable caseine, see the preceding note.

NOTE (10), p. 64.

AMOUNT OF MATTER SOLUBLE IN ALCOHOL IN THE SOLID EXCREMENTS OF THE HORSE AND COW. (WILL.*)

18·3 grammes of dried horse-dung lost, by the action of alcohol, 0·995 gramme. The residue, when dry, had the appearance of saw-dust, after it has been deprived, by boiling, of all soluble matter.

14·98 grammes of dry cow-dung lost, by the same treatment, 0·625 gramme.

NOTE (11), p. 70.

COMPOSITION OF STARCH. *a*

	Calculated $C_{12} H_{10} O_{10}$.	Strecker.*			
		From Peas.	From Lentils.	From Beans.	From Buckwheat.
Carbon	44·91	44·33	44·46	44·16	44·23
Hydrogen	6·11	6·57	6·54	6·69	6·40
Oxygen	48·98	49·09	49·00	49·15	49·37

Strecker.*

	From maize.	From horse-chesnuts.	From wheat.	From rye.
Carbon	44·27	44·44	44.26	44·16
Hydrogen...	6·67	6·47	6·70	6·64
Oxygen ...	49·06	49·08	49·04	49.20

Strecker.*

	From rice.	From dahlia roots.	From unripe apples.	From unripe pears.
Carbon	44·69	44·13	44·10	44·14
Hydrogen ...	6·36	6·56	6·57	6·75
Oxygen......	48·95	49·31	49.33	49·11

	From potatoes.		From arrow-root.	From yams.*a*
	Berzelius.	Gay Lussac & Thenard.	Prout.	Ortigosa.*
Carbon ...	44·250	43·55	44·40	44·2
Hydrogen	6·674	6·77	6·18	6·5
Oxygen...	49·076	49·68	49·42	49.3

a The starch employed for the analyses, made by Strecker and Ortigosa, was prepared from the chemical laboratory at Giessen, from the respective seeds, bulbs, and fruits.

NOTE (12), p. 71.

COMPOSITION OF GRAPE SUGAR. (STARCH SUGAR.)

	From grapes.*a* De Saussure.	From starch.*b*	From honey.*c* Prout.	Calculated. $C_{12}H_{14}O_{14}$
Carbon ...	36·71	37·29	36·36	36·80
Hydrogen	6·78	6·84	7·09	7·01
Oxygen...	56·51	55.87	56·55	56·19

a Ann. de Chimie, XI., 381.
b Ann. of Philosophy, VI., 426.
c Philosoph. Trans. 1827, 373.

NOTE (13), p. 72.

COMPOSITION OF SUGAR OF MILK.

	Gay Lussac and Thenard.	Prout.	Brunn.	Berzelius.	Liebig.*	Calculated. $C_{12} H_{12} O_{12}$.
Carbon ...	38·825	40·00	40·437	39·474	40·00	40·46
Hydrogen	7·341	6·66	6·711	7·167	6·73	6·61
Oxygen ...	53·834	53·34	52·852	53,359	53·27	52·93

NOTE (14), p. 72.

COMPOSITION OF GUM.

	Gay Lussac and Thenard.	Goebel.	Berzelius.	Calculated. $C_{12} H_{11} O_{11}$.
Carbon	42·23	42·2	42·682	42·58
Hydrogen	6·93	6·6	6·374	6·37
Oxygen.........	50·84	51·2	50·944	51·05

NOTE (15), p. 74.

ANALYSIS OF OATS (Boussingault). *a*

100 parts of oats contain of dry matter .... 84·9

Ditto   water .... 17·1

_____

100·0

100 parts of oats dried at 212° = 117·7 parts dried at the ordinary temperature, contain

Carbon	50·7
Hydrogen	6·4
Oxygen	36·7
Nitrogen	2·2
Ashes	4·0

_____

100·0

Water..................... 17·7

_____

Oats dried in the air ... 117·7 contain, in 100 parts, 1·867 of nitrogen.

a Ann. de Chimie et de Phys., LXXI., 130.

ANALYSIS OF HAY.

100 parts of hay dried in the air contain 86 of dry matter,

14 of water.

——

100

100 parts of hay dried at 212° = 116·2 parts dried in air, contain

Carbon..................	45·8
Hydrogen..............	5·0
Oxygen	38·7
Nitrogen	1·5
Ashes	9·0

——

100·0

16.2 water,

——

116·2 hay dried in the air.

100·0 of hay dried at the ordinary temperature contain 1·29 of nitrogen.

240 oz. of such hay = 15 lbs. contain ... 3·095 oz. of nitrogen.

72 oz. of oats = 4½ lbs. contain ... 1·34   ditto

Total ...... 4·435   ditto

———

NOTE (16), a, p. 77.

AMOUNT OF CARBON IN FLESH AND IN STARCH.

100 parts of starch contain 44 of carbon ; therefore, 64 oz. (4 lbs.) contain 28·16 oz. of carbon.

100 parts of fresh meat contain 13·6 of carbon (see Note III.); hence 240 oz. (15 lbs.) contain 32·64 oz. of carbon.‡

‡ By an error in calculation in the original, the amount of carbon in 15 lbs. of meat is stated to be 27·64 oz. It follows, that the carbon of 4 lbs. of starch is not equal, as stated in the text, to that of 15 lbs. of flesh, but to that of 13 lbs. This difference, however, is not sufficient to affect the argument at p. 84.— EDITOR.

NOTE (16), b, p. 84.

COMPOSITION OF

	Hog's Lard.	Mutton fat. Chevreul. a	Human Fat.
Carbon............	79·098	78·996	79.000
Hydrogen	11·146	11·700	11·416
Oxygen	9·756	9·304	9·584

a Recherches Chim., sur les corps gras. Paris. 1823.

NOTE (17), p. 84.

COMPOSITION OF CANE SUGAR.

	Berzelius.	Prout.	According to W. Crum.	Liebig.*	Gay Lussac & Thenard.	Calculated. $C_{12} H_{11} O_{11}$.
Carbon......	42·225	42·86	42·14	42·301	42·47	42·58
Hydrogen...	6·600	6·35	6·42	6·384	6·90	6·37
Oxygen ...	51·175	50·79	51·44	51·315	50·63	51·05

For the composition of gum and of starch, see Notes (14) and (11).

NOTE (18), p. 85.

COMPOSITION OF CHOLESTERINE.

	Chevreul. a	According to Couerbe. b	Marchand.	Calculated. $C_{36} H_{32} O$.
Carbon ...	85·095	84·895	84·90	84·641
Hydrogen	11·880	12·099	12·00	12·282
Oxygen ...	3·025	3·006	3·10	3·077

a Recherches sur les corps gras, p. 185.
b Ann. de Ch. et de Phys. LVI., p. 164.

NOTE (19), p. 87.

THE PRODUCTION OF WAX FROM SUGAR. *a*

As soon as the bees have filled their stomach, or what is called the honey bladder, with honey, and cannot deposit it for want of cells, the honey passes gradually in large quantity into the intestinal canal, where it is digested. The greater part is expelled as excrement; the rest enters the fluids of the bee. In consequence of this great flow of juices a fatty substance is produced, which oozes out on the eight spots formerly mentioned, which occur on the four lower scales of the abdominal rings, and soon hardens into laminæ of wax. On the other hand, when the bees can deposit their honey, only so much enters the intestinal canal as is necessary for their support. The honey bladder need not be filled with honey longer than forty hours in order to bring to maturity, on the eight spots, eight laminæ of wax, so that the latter fall off. I made the experiment of giving to bees, which I had enclosed in a box with their queen about the end of September, dissolved sugar-candy instead of honey. Out of this food laminæ of wax were formed; but these would not separate and fall off readily, so that the mass, which continued to ooze out, remained, in most of the bees, hanging to the upper lamina; and the laminæ of wax became as thick as four under ordinary circumstances. The abdominal scales of the bees were, by means of the wax, distinctly raised, so that the waxen laminæ projected between them. On examination, I

---

a From F. W. Gundlach's Natural History of Bees, p. 115. Cassel, 1842. We are acquainted with no more beautiful or convincing proof of the formation of fatty matter from sugar than the following process of the manufacture of wax by the bee as taken from observation.

APPENDIX.

found that these thick laminæ, which under the micro-
scope exhibited several lamellæ, had a sloping surface
downwards near the head, and upwards in the vicinity of
the tail. The first waxen lamina, therefore, must have
been pushed downwards by the second, because, where
the abdominal scales are attached to the skin, there is no
space for two laminæ, the second by the third, and thus
the inclined surfaces on the sides of the thick laminæ had
been produced. I saw distinctly from this, that the first-
formed laminæ are detached by those which follow. The
sugar had been converted into wax by the bees, but it
would seem that there was some imperfection in the process, as the laminæ did not fall off, but adhered to the
succeeding ones.

In order to produce wax in the manner described, the
bees require no pollen, but only honey. I have placed,
even in October, bees in an empty hive, and fed them
with honey; they soon formed comb, although the weather was such that they could not leave the hive. I cannot, therefore, believe that pollen furnishes food for the
bees, but I think they only swallow it in order, by mixing
it with honey and water, to prepare the liquid food for the
grubs. Besides, bees often starve in April, when their
stock of honey is consumed, and when they can obtain in
the fields abundance of pollen, but no honey. When
pressed by hunger they tear the nymphæ out of the cells,
and gnaw them in order to support life by the sweet
juice which they contain. But, if in this condition they are
not artificially fed, or if the fields do not soon yield their
proper food, they die in the course of a few days. Now,
if the pollen were really nourishment for bees, they ought
to be able to support life on it, mixed with water.

Bees never build honeycomb unless they have a queen,
or are provided with young out of which they can educate

a queen. But if bees be shut up in a hive without a queen, and fed with honey, we can perceive in forty-eight hours that they have laminæ of wax on their scales, and that some have even separated. The building of cells is therefore voluntary, and dependant on certain conditions, but the oozing out of wax is involuntary.

One might suppose that a large proportion of these laminæ must be lost, since the bees may allow them to fall off, out of the hive as well as in it; but the Creator has wisely provided against such a loss. If we give to bees engaged in building cells honey in a flat dish, and cover the dish with perforated paper, that the bees may not be entangled in the honey, we shall find, after a day, that the honey has disappeared, and that a large number of laminæ are lying on the paper. It would appear as if the bees, which have carried off the honey, had let fall the scales; but it is not so. For, if above the paper we lay two small rods, and on these a board, overhanging the dish on every side, so that the bees can creep under the board and obtain the honey, we shall find next day the honey gone, but no laminæ on the paper; while laminæ will be found in abundance on the board above. The bees, therefore, which go for and bring the honey, do not let fall the laminæ of wax, but only those bees which remain hanging to the top of the hive. Repeated experiments of this kind have convinced me that the bees, as soon as their laminæ of wax are mature, return to the hive and remain at rest, just as caterpillars do, when about to change. In a swarm that is actively employed in building we may see thousands of bees hanging idly at the top of the hive. These are all bees whose laminæ of wax are about to separate. When they have fallen off, the activity of the bee revives, and its place is occupied for the same purpose by another.

(From page 28 of the same work.) In order to ascertain how much honey bees require to form wax, and how often, in a swarm engaged in building, the laminæ attain maturity and fall off, I made the following experiment, which appears to me not uninteresting.

On the 29th of August, of this year (1841), at a time when the bees could obtain in this district no farther supply of honey from the fields, I emptied a small hive, placed the bees in a small wooden hive, having first selected the queen bee, and shut her up in a box, furnished with wires, which I placed in the only door of the hive, so that no embryos could enter the cells. I then placed the hive in a window, that I might be able to watch it.

At 6 p.m. I gave the bees 6 oz. of honey run from the closed cells, which had thus the exact consistence of freshly made honey. This had disappeared next morning. In the evening of the 30th I gave the bees 6 oz. more, which, in like manner, was removed by the next morning; but already some laminæ of wax were seen lying on the paper with which the honey was covered. On the 31st August and the 1st September the bees had in the evening 10 oz., and on the 3rd of September in the evening 7 oz.; in all, therefore, 1 lb. 13 oz. of honey, which had run cold out of cells which the bees had already closed. On the 5th of September I stupified the bees, by means of puff-ball, and counted them. Their number was 2,765, and they weighed 10 oz. I next weighed the hive, the combs of which were well filled with honey, but the cells not yet closed; noted the weight, and then allowed the honey to be carried off by a strong swarm of bees. This was completely effected in a few hours. I now weighed it a second time, and found it 12 oz. lighter; consequently the bees still had in the hive 12 oz. of the 29 oz. of honey given to them. I next extracted the combs, and found that their weight was ⅝ of an ounce. I then placed the bees in another box, provided with empty combs, and fed them with the same honey as before. In the first few days they lost daily rather more than 1 oz. in weight, and afterwards half an ounce daily, which was owing to the circumstance, that from the digestion of so much honey, their intestinal canal was loaded with excrements; for 1,170 bees, in autumn, when they have been but a short time confined to the hive, weigh 4 oz.; consequently 2,765 bees should weigh 9 oz. But they actually weighed 10 oz., and therefore had within them 1 oz. of excrement, for their honey bladders were empty. During the night the weight of the box did not diminish at all, because the small quantity of honey the bees had deposited in the cells, having already the proper consistence, could not lose weight by evaporation, and because the bees could not then get rid of their excrements. For this reason, the loss of weight occurred always during the day.

If, then, the bees, in seven days, required 3½ oz. of honey to support and nourish their bodies, they must have consumed 13½ oz. of honey in forming ⅝ of an ounce of wax; and consequently, to form 1 lb. of wax, 20 lbs. of honey are required. This is the reason why the strongest swarms in the best honey seasons, when other hives, that have no occasion to build, often gain in one day 3 or 4 lbs. in weight, hardly become heavier, although their activity is boundless. All that they gain is expended in making wax. This is a hint for those who keep bees, to limit the building of comb. Cnauf has already recommended this, although he was not acquainted with the true relations of the subject. From 1 oz. of wax, bees can build cells enough to contain 1 lb. of honey.

100 laminæ of wax weigh 0·024 gramme (rather more than ⅓ of a grain), consequently, 1 kilogramme (= 15,360

grains) will contain 4,166,666 laminæ. Hence, $\frac{5}{6}$ of an
ounce will contain 81,367 laminæ. Now this quantity
was produced by 2,765 bees in six days; so that the bee
requires for the formation of its 8 laminæ (one crop)
about thirty-eight hours, which agrees very well with my
observations.

The laminæ, when formed, are as white as bleached
wax. The cells also, at first, are quite white, but they are
coloured yellow by the honey, and still more by the
pollen. When the cold weather comes on, the bees
retire to the hive under the honey, and live on the stock
they have accumulated.

P. 54. Many believe that bees are hybernating
animals; but this opinion is quite erroneous. They are
lively throughout the winter; and the hive is always
warm in consequence of the heat which they generate.
The more numerous the bees in a hive, the more heat is
developed; and hence strong hives can resist the most
intense cold. It once happened that I forgot to remove
from the door, which was unusually large, of a hive in
in winter, a perforated plate of tinned iron, which I had
fastened over the opening to diminish the heat in July;
and yet this hive came well through the winter, although
the cold was very severe, having been for several days
so low as 0°. But I had added to this hive the bees of
two other hives! When the cold is very intense, the bees
begin to hum. By this means respiration is accelerated
and the developement of heat increased. If, in summer,
bees without a queen are shut up in a glass box, they
become uneasy and begin to hum. So much heat is by
this means developed, that the plates of glass become
quite hot. If the door be not opened in this case, or if
air be not admitted, and if the glass be not cooled by the
aid of water, the bees are soon suffocated.

COMPOSITION OF BEES' WAX.

	Gay Lussac. & Thénard.a	De Saussure.b	Oppermann.c	Ettling.d	Hess.e	Calculated $C_{20}H_{20}O$.
Carbon ...	81·784	81·607	81·291	81·15	81·52	81·38
Hydrogen	12·672	13·859	14·073	13·75	13·23	13·28
Oxygen ...	5·544	4·534	4·636	5·09	5·25	5·34

a Traité de Chimie, par Thénard, 6me· Ed., IV., 477.
b Ann. de Ch. et de Phys., XIII., 310.
c Ibid. XLIX., 224.
d Annal. der Pharm., II., 267.
e Ibid. XXVII., 6.

NOTE (21) a, p. 104.

COMPOSITION OF HYDRATED CYANURIC ACID, OF HYDRATED CYANIC ACID, AND OF CYAMELIDE, IN 100 PARTS, ACCORDING TO THE ANALYSIS OF WOHLER AND LIEBIG.* a

	Cyanuric acid, cyanic acid, cyamelide.
Carbon	28·19
Hydrogen	2·30
Nitrogen	32·63
Oxygen	36·87

a Poggendorff's Annalen, XX., 375 et seq.

NOTE (21) b, p. 104.

COMPOSITION OF ALDEHYDE, METALDEHYDE, AND ELALDEHYDE. a

	Aldehyde. Liebig.*	Metaldehyde.	Elaldehyde. Fehling.*		Calculated $C_4H_4O_2$.
Carbon	55·024	54·511	54·620	54·467	55·024
Hydrogen ...	8·983	9·054	9·248	9·075	8·983
Oxygen	35·993	36·435	36·132	36·458	35·993

a Ann. der Pharm., XIV., 142, and XXVII., 319.

x 2

NOTE (22), p. 105.

COMPOSITION OF PROTEINE.

	From the crystalline lens.	From albumen. Scherer. a	From fibrine.
Carbon	55·300	55·160	54·848
Hydrogen	6·940	7·055	6·959
Nitrogen	16·216	15·966	15·847
Oxygen.........	21·544	21·819	22·346

	Scherer. a From hair.		From horn.		Calculated $C_{48}H_{36}N_6O_{14}$.
Carbon......	54·746	55·150	55·408	54·291	55·742
Hydrogen	7·129	7·197	7·238	7·082	6·827
Nitrogen...	15·727	15·727	15·593	15·593	16·143
Oxygen ...	22·398	21·926	21·761	23·034	21·228

a Ann. der Chim. und Pharm., XL., 43.

	From vegetable albumen.	From fibrine. Mulder. a	From albumen.	From cheese.
Carbon	54·99	55·44	55·30	55·159
Hydrogen ...	6·87	6·95	6·94	7·176
Nitrogen......	15·66	16·05	16·02	15·857
Oxygen	22·48	21·56	21·74	21·808

a Ann. der Pharm., XXVIII., 75.

NOTE (23), p. 107.

COMPOSITION OF THE ALBUMEN OF THE YOLK AND OF THE WHITE OF THE EGG. a

	From the yolk. Jones.		From the white. Scherer.
	I.	II.	
Carbon.........	53·72	53·45	55·000
Hydrogen ...	7·55	7·66	7·073
Nitrogen	13·60	13·34	15·920
Oxygen Sulphur Phosphorus }	25·13	25·55	22·007

a Ann. der Chem. und Pharm., XL., 36, ibid. 67.

NOTE (24), p. 111.

COMPOSITION OF LACTIC ACID.

$C_6H_5O_5.$

Carbon..................	44·90
Hydrogen...............	6·11
Oxygen	48·99

NOTE (25), p. 115.

GAS FROM THE ABDOMEN OF COWS AFTER EATING CLOVER TO EXCESS, OBTAINED BY PUNCTURE.

a Examined by Lameyran and Frémy. *b* By Vogel.
c By Pflüger.

	Air.	Carbonic acid.	Inflammable gas.	Sulphuretted hydrogen.
a	5	5	—	15	80 Vol. in 100 Vol.
b	25	—	27	48	—
c	—	—	60	40	—
c	—	—	20	80	—

NOTE (26), p. 118.

MAGENDIE FOUND IN THE STOMACH AND INTESTINES OF EXECUTED CRIMINALS:

a In the case of an individual who had taken food in moderation one hour previous to death ; *b,* in the case of one who had done so two hours previously ; and *c,* in the case of a third, who had done so four hours previous to execution.

		Oxygen.	Nitrogen.	Carbonic acid.	Inflammable gas.
	From the stomach.............	11·00 Vol.	71·45	14·00	3·55
a	— small intestines ...	00·00	20·03	24·39	55·53
	— large intestines ...	00·00	51·03	43·50	5·47
	From the stomach.............	00·00	00·00	00·00	00·00
b	— small intestines ...	00·00	8·85	40·00	51·15
	— large intestines ...	00·00	18·40	70·00	11·60
	From the stomach.............	00·00	00·00	00·00	00·00
c	— small intestines ...	00·00	66·60	25·00	8·40
	— large intestines ...	00·00	45·96	42·86	11·18

100 Volumes of the gas contained.

NOTE (27), referred to in NOTE (7), p. 43.

COMPOSITION OF ANIMAL ALBUMEN AND FIBRINE, AND OF THE DIFFERENT TISSUES OF THE BODY.

1. ALBUMEN.

	From the serum of blood. Scherer.[*] a				From eggs.	From yolk of egg. Jones.[*] b	
	I.	II.	III.	IV.	V.	VI.
Carbon.........	53·850	55·461	55·097	55·000	53·72	53·45
Hydrogen ...	6·983	7·201	6·880	7·073	7·55	7·66
Nitrogen	15·673	15·673	15·681	15·920	13·60	13·34
Oxygen ⎫ Sulphur ⎬ Phosphorus ⎭	23·494	21·655	22·342	22·007	25·13	25·55

a Ann. der Chem. und Pharm., XL., 36.
b Ibid. 67.

	Jones.[*]	Scherer.[*]				
	From albumen of brain.	From hydrocele.	From congestive abscess.	From pus.		From fluid of dropsy.
	VII.	VIII.	IX.	X.	XI.	XII.
Carbon	55·50	54·921	54·757	54·663	54·101	54·302
Hydrogen ...	7·19	7·077	7·171	7·022	6·947	7·176
Nitrogen ...	16·31	15·465	15·848	15·839	15·660	15·717
Oxygen ⎫ Sulphur ⎬ Phosphorus ⎭	21·00	22·537	22·224	22·476	23·292	22·805

	Mulder. a
Carbon	54·84
Hydrogen......	7·09
Nitrogen	15·83
Oxygen.......................	21·23
Sulphur	0·68
Phosphorus	0·33

a Ann. der Pharm. XXVIII., 74.

2. FIBRINE.

Scherer.* a

	I.	II.	III.	IV.	V.	VI.	VII.
Carbon	53·671	54·454	55·002	54·967	53·571	54·686	54·844
Hydrogen ...	6·878	7·069	7·216	6·867	6·895	6·835	7·219
Nitrogen ...	15·763	15·762	15·817	15·913	15·720	15·720	16·065
Oxygen ⎫ Sulphur ⎬ Phosphorus ⎭	23·688	22·715	21·965	22·244	23·814	22·759	21·872

a Ann. der Chem. und Pharm., XL., 33.

	Mulder.*a*
Carbon	54·56
Hydrogen	6·90
Nitrogen	15·72
Oxygen........................	22·13
Sulphur	0·33
Phosphorus	0·36

a Ann. der Pharm., XXVIII., 74.

3. GELATINOUS TISSUES.

Scherer.* a

	Isinglass.	Tendons of the calf's foot.			Tunica sclerotica.	Calculated. $C_{13}H_{41}N7\frac{1}{2}O_{18}$
Carbon ...	50·557	49.563	50·960	50·774	50·995	50·207
Hydrogen	6·903	7·148	7·188	7·152	7·075	7·001
Nitrogen	18·790	18·470	18·320	18·320	18·723	18·170
Oxygen...	23·750	24·819	23·532	23·754	23·207	24·622

a Ann. der Chem. und Pharm., XL., 46.

	Mulder.	
Carbon................	50·048	50·048
Hydrogen	6·477	6·643
Nitrogen	18·350	18·388
Oxygen	25·125	24·921

4. Tissues containing Chondrine.

	Scherer.[*] a				
	Cartilages of the ribs of the calf.		Cornea.	Calculated. $C_{48}H_{40}N_6 O_{20}$.	Mulder.
Carbon	49·496	50·895	49·522	50·745	50·607
Hydrogen...	7·133	6·962	7·097	6·904	6·578
Nitrogen ...	14·908	14·908	14·399	14·692	14·437
Oxygen ...	28·463	27·235	28·982	27·659	28·378

a Ann. der Chem. und Pharm., XL., 49.

5. Composition of the middle Membrane of Arteries.

	Scherer.[*] a		Calculated.
	I.	II.	$C_{48}H_{38}N_6 O_{15}$.
Carbon	53·750	53·393	53·91
Hydrogen	7·079	6·973	6·96
Nitrogen	15·360	15·360	15·60
Oxygen	23·811	24·274	23·53

a Ann. der Chem. und Pharm., XL., 51.

6. Composition of Horny Tissues.

	Scherer.[*] a					
	External skin of the sole of the foot.		Hair of the beard.		Hair of the head.	
				Fair.	Brown.	Black.
Carbon ...	51·036	50·752	51·529	50·652	49·345	50·622	49·935
Hydrogen	6·801	6·761	6·687	6·769	6·576	6·613	6·631
Nitrogen	17·225	17·225	17·936	17·936	17·936	17·936	17·936
Oxygen ... Sulphur ... }	24·938	25·262	23·848	24·643	26·143	24·829	25·498

	Scherer.[*]						Calculated.
	Buffalo horn.				Nails.	Wool.	$C_{48}H_{39}N_7O_{17}$.
Carbon ...	51·990	51·162	51·620	51·540	51·089	50·653	51·718
Hydrogen	6·717	6·597	6·754	6·779	6·824	7·029	6·860
Nitrogen	17·284	17·284	17·284	17·284	16·901	17·710	17·469
Oxygen ... Sulphur ... }	24·009	24·957	24·342	24·397	25·186	24·608	23·953

a Ann. der Chem. und Pharm., XL., 53.

The composition of the membrane lining the interior of the shell of the egg approaches closely to that of horn. According to Scherer, it contains

	Scherer.[*]a
Carbon	50·674
Hydrogen	6·608
Nitrogen.....................	16·761
Oxygen ⎱	25·957
Sulphur ⎰	

a Ann. der Chem. und Pharm., XL., 60.

The composition of feathers is also nearly the same as that of horn.

	Scherer.[*]a		Calculated.
	Beard of the feather.	Quill of the feather.	$C_{48}H_{39}N_7O_{16}$.
Carbon	50·434	52·427	52·457
Hydrogen............	7·110	7·213	6·958
Nitrogen	17·682	17·893	17·719
Oxygen	24·774	22·467	22·866

The analysis here given of the beard of feathers agrees closely with that of horn, while that of the quill is more accurately represented by the attached formula, which differs from that of horn by 1 eq. of oxygen only.

a Ann. der Chem. und Pharm. XL., 61.

7. Composition of the Pigmentum nigrum Oculi,

	Scherer.[*]a		
Carbon	58·273	58·672	57·908
Hydrogen...	5·973	5·962	5·817
Nitrogen ...	13·768	13·768	13·768
Oxygen ...	21·986	21·598	22·507

a Ann. der Chem. und Pharm., XL., 63.

NOTE (28), p. 133.

According to the analyses of Playfair and Bœckmann,

0·452 parts of dry muscular flesh gave 0·836 of carbonic acid.

0·407 .... 0·279 of water.

0·242 .... 0·450 of carbonic acid and 0·164 of water.

0·191 .... 0·360 .... 0·130

0·305 of dried blood gave 0·575 carbonic acid and 0·202 of water.

0·214 .... 0·402 .... 0·138

1·471 of dried blood, when calcined, left 0·065 of ashes = 4·42 per cent.

     The dried flesh was found to contain   of ashes   4·23 per cent.

     The nitrogen was found to be to the carbon as 1 to 8 in equivalents.

Hence

	Flesh (beef).		Ox-blood.		Blood.
	Playfair.	Bœckmann.	Playfair.	Bœckmann.	Mean of 2 analyses.
Carbon......	51·83	51·89	51·95	51·96	51·96
Hydrogen	7·57	7·59	7·17	7·33	7·25
Nitrogen ...	15·01	15·05	15·07	15·08	15·07
Oxygen ...	21·37	21·24	21·39	21·21	21·30
Ashes	4·23	4·23	4·42	4·42	4·42

Deducting the ashes, or inorganic matter, the composition of the organic part is,

Carbon	54·12	54·18	54·19	54·20
Hydrogen ...	7·89	7·93	7·48	7·65
Nitrogen ...	15·67	15·71	15·72	15·73
Oxygen......	22·32	22·18	22·31	22·12

This corresponds to the formula

C_{48}	....	54·62
H_{39}	....	7·24
N_6	....	15·81
O_{15}	....	22·33

NOTE (29), p. 134.

COMPOSITION OF CHOLEIC ACID. a

	Demarçay.	Dumas.	Calculated $C_{76}H_{66}N_2O_{22}$.
Carbon	63·707	63·5	63·24
Hydrogen ...	8·821	9·3	8·97
Nitrogen ...	3·255	3·3	3·86
Oxygen	24·217	23·9	23·95

a Ann. der Pharm., XXVII., 284 and 293.

NOTE (30), p. 135.

COMPOSITION OF TAURINE AND OF CHOLOIDIC ACID.

1. TAURINE. a

	Demarçay.*	Dumas.	Calculated $C_4H_7NO_{10}$.
Carbon	19·24	19·26	19·48
Hydrogen ...	5·78	5·66	5·57
Nitrogen ...	11·29	11·19	11·27
Oxygen	63·69	63·89	63·68

a Ann. der Pharm., XXVII., 287 and 292.

2. CHOLOIDIC ACID. a

	Demarçay.*		Dumas.	Calculated $C_{36}H_{56}O_{12}$.
	I.	II.		
Carbon	73·301	73·522	73·3	74·4
Hydrogen ...	9·511	9·577	9·7	9·4
Oxygen ...	17·188	16·901	17·0	16·2

a Ann. der Pharm., XXVII., 289 and 293.

In reference to the researches of Demarçay on the bile I would make the following observations.

The matter to which I have given the name of *choleic acid* is the bile itself separated from the inorganic constituents (salts, soda, &c.) which it contains. By the action of subacetate of lead aided by ammonia, all the organic constituents of the bile are made to unite with oxide of lead, with which they form an insoluble, resinous precipitate. The substance here combined with oxide of lead contains all the carbon and nitrogen of the bile. The substance which I have named *choloidic acid* is that which is obtained, when the bile, purified by alcohol from the substances insoluble in that fluid, is boiled for some time with an excess of muriatic acid. It contains all the carbon and hydrogen of the bile, except those portions which have separated in the form of taurine and ammonia. The *cholic acid* contains the elements of bile, *minus* those of carbonate of ammonia.

These three compounds, therefore, contain the products of the metamorphosis of the entire bile ; their formulæ express the amount of the elements of the constituents of the bile. No one of them exists ready formed in the bile in the shape in which we obtain it ; their elements are combined in a different way from that in which they were united in the bile ; but the way in which these elements are arranged has not the slightest inference on the determination by analysis of the relative proportions of the elements. In the formulæ themselves, therefore, is involved no hypothesis ; they are simply expressions of the results of analysis. It signifies nothing that the choleic or choloidic acids may be composed of several compounds united together. No matter how many such they may contain, the relative proportions of all the elements taken together is expressed by the formula which is derived from the analysis.

The study of the products which are produced from the

bile by the action of the atmosphere, or of chemical re-
agents, may be of importance in reference to certain pa-
thological conditions; but except as concerns the general
character of the bile, the knowledge of these products
is of no value to the physiologist; it is only a burthen
which impedes his progress. It cannot be maintained of
any one of the 38 or 40 substances, into which the bile
has been divided or split up, that it exists ready formed
in the healthy secretion; on the contrary, we know with
certainty that most of them are mere products of the action of the re-agents which are made to act on the bile.

The bile contains soda; but it is a most remarkable
and singular compound of soda. When we cause that
part of the bile which dissolves in alcohol (which contains
nearly all the organic part) to combine with oxide of lead,
thus separating the soda, and then remove the oxide of
lead, we obtain a substance, choleic acid, which, when
placed in contact with soda, forms a compound similar to
bile in its taste; but it is no longer bile; for bile may be
mixed with organic acids, nay, even with dilute mineral
acids, without becoming turbid or yielding a precipitate;
while the new compound, choleate of soda, is decomposed
by the feeblest acids, the whole of the choleic acid being
separated. Hence, bile cannot be considered, in any
sense, as choleate of soda. Further, it may be asked, in
what form are the cholesterine, and stearic, and margaric
acids, which are found in bile, contained in that fluid?
Cholesterine is insoluble in water, and not saponifiable by
alkalies; and if the two fatty acids just named were
really present in the bile as soaps of soda, they would be
instantly separated by other acids. Yet diluted acids cause
no such separation of stearic and margaric acids in bile.

It is possible that, in the course of new and repeated
investigations, the composition of the substances obtained

from bile may be found different from that which has been given in our analytical developement of this subject. But this, if it should happen, can have but little effect on our formulæ; if the relative proportions of carbon and nitrogen be not changed, the differences will be confined to the proportions of oxygen and hydrogen. In that case it will be necessary for the developement of our views in formulæ, only to assume that more water and oxygen, or less water and oxygen, have taken a share in the metamorphosis of the tissues; but the truth of the developement of the process itself will not be by this means affected.

NOTE (31), p. 135.

COMPOSITION OF CHOLIC ACID. *a*

	Dumas.	Calculated $C_{74}H_{60}O_{18}$.
Carbon	68·5	68·9
Hydrogen	9·7	9·2
Oxygen	21·8	21·9

a Ann. der Pharm., XXVII., 295.

NOTE (32), p. 137.

COMPOSITION OF THE CHIEF CONSTITUENTS OF THE URINE OF MEN AND ANIMALS.

1. URIC ACID.

	Liebig.* *a*	Mitscherlich. *b*	Calculated $C_{10}H_4N_4O_6$.
Carbon......	36·083	35·82	36·00
Hydrogen...	2·441	2·38	23·6
Nitrogen ...	33·361	34·60	33·37
Oxygen ...	28·126	27·20	28·27

a Ann. der Pharm., X., 47.
b Poggendorff's Ann., XXXIII., 335.

2. ALLOXAN. *a*

A PRODUCT OF THE OXIDATION OF URIC ACID.

	Wöhler and Liebig. *		Calculated C₃H₄N₂O₁₀.
Carbon......	30·38	30·18	30·34
Hydrogen...	2·57	2·48	2·47
Nitrogen ...	17·96	17·96	17·55
Oxygen ...	49·09	49·38	49·64

a Ann. der Pharm., XXVI., 260.

3. UREA.

	Prout. *a*	Wöhler and Liebig. *b*	Calculated C₂H₄N₂O₂.
Carbon......	19·99	20·02	20·192
Hydrogen...	6·65	6·71	6·595
Nitrogen ...	46·65	46·73	46·782
Oxygen	26·63	26·54	26·425

a Thomson's Annals, XI., 352.
b Poggend. Ann., XX., 375.

4. CRYSTALLIZED HIPPURIC ACID.

	Liebig.* *a*	Dumas. *b*	Mitscherlich. *c*	Calculated C₁₈H₃NO₆.
Carbon......	60·742	60·5	60·63	60·76
Hydrogen...	4·959	4·9	4·98	4·92
Nitrogen ...	7·816	7·7	7·90	7·82
Oxygen ...	26·483	26·9	26·49	26·50

a Ann. der Pharm., XII., 20.
b Ann. de Ch. et de Phys., LVII., 327.
c Poggend. Ann., XXXIII., 335.

5. ALLANTOINE. *a*

	Wöhler and Liebig. *	Calculated C₈H₆N₄O₆.
Carbon	30·60	30·66
Hydrogen	3·83	3·75
Nitrogen	35·45	35·50
Oxygen	30·12	30·09

a Ann. der Pharm., XXVI., 215.

6. Uric or Xanthic Oxide. a

	Wöhler and Liebig.*	Calculated $C_5 H_2 N_2 O_2$.
Carbon	39·28	39·86
Hydrogen	2·95	2·60
Nitrogen	36·35	37·72
Oxygen............	21·24	20·82

a Ann. der Pharm., XXVI., 344.

7. Cystic Oxide. a

	Thaulow.*	Calculated $C_6 H_6 NO_4 S_2$.
Carbon	30·01	30·31
Hydrogen	5·10	4·94
Nitrogen	11·00	11·70
Oxygen............	28·38	26·47
Sulphur............	25·51	26·58

a Ann. der Pharm. XXVII., 200.

The cystic oxide is distinguished from all the other concretions occurring in the urinary bladder by the sulphur it contains. It can be shewn with certainty, that the sulphur is present neither in the oxidised state, nor in combination with cyanogen; and in regard to its origin the remark is not without interest, that four atoms of cystic oxide contain the elements of uric acid, benzoic acid, sulphuretted hydrogen, and water; all of which are substances, the occurrence of which in the body is beyond all doubt.

1 atom uric acid ...	$C_{10} N_4 H_4 O_6$
1 atom benzoic acid	$C_{14} \quad H_5 O_3$
8 atoms sulphuret-⎱ ted hydrogen ...⎰	$H_8 \quad S_8$
7 atoms water	$H_7 O_7$

4 atoms cystic oxide $= C_{24} N_4 H_{24} O_{16} S_8 = 4 (C_6 NH_6 O_4 S_2)$.

An excellent method of detecting the presence of cystic oxide in calculi or gravel is the following :

The calculus is dissolved in a strong solution of caustic potash, and to the solution is added so much of a solution of acetate of lead, that all the oxide of lead is retained in solution. When this mixture is boiled there is formed a black precipitate of sulphuret of lead, which gives to the liquid the aspect of ink. Abundance of ammonia is also disengaged; and the alkaline fluid is found to contain, among other products, oxalic acid.

NOTE (33), p. 137.

COMPOSITION OF OXALIC, OXALURIC, AND PARABANIC ACIDS.

1. OXALIC ACID (hydrated).

	Gay Lussac & Thénard.	Berthollet.	Calculated $C_2 O_3 + H O$
Carbon.........	26·566	25·13	26·66
Hydrogen ...	2·745	3·09	2·22
Oxygen	70·689	71·78	71·12

2. OXALURIC ACID. a

	Wöhler and Liebig.*		Calculated $C_6 H_4 N_2 O_3$
Carbon	27·600	27·318	27·59
Hydrogen	3·122	3·072	3·00
Nitrogen	21·218	21·218	21·29
Oxygen	48·060	48·392	48·12

a Ann. der Pharm., XXVI., 289.

3. PARABANIC ACID. a

Wöhler and Liebig.*

			Calculated
Carbon.........	31·95	31·940	31·91
Hydrogen......	2·09	1·876	1·73
Nitrogen	24·66	24·650	24·62
Oxygen	41·30	41·534	41·74

a Ann. der Pharm., XXVI., 286.

NOTE (34), p. 138.

COMPOSITION OF ROASTED FLESH.

(1.) 0·307 of flesh gave 0·584 of carbonic acid and 0·206 of water.
(2.) 0·255   do.   0·485   do.   0·181   do.
(3.) 0·179   do.   0·340   do.   0·125   do.

Hence—

	Flesh of roedeer (1). Bœckmann.[*]	Flesh of beef (2). Playfair.[*]	Flesh of veal (3).
Carbon	52·60	52·590	52·52
Hydrogen	7·45	7·886	7·87
Nitrogen	15·23	15·214	14·70
Oxygen } Ashes }	24·72	24·310	24·91

NOTE (35), p. 142.

The formula $C_{108}H_{84}N_{18}O_{40}$, or $C_{54}H_{42}N_9O_{20}$, gives, when reduced to 100 parts,

$$C_{54} \quad 50·07$$
$$H_{42} \quad 6·35$$
$$N_9 \quad 19·32$$
$$O_{20} \quad 24·26$$

Compare this with the composition of gelatine, as given in Note (27).

NOTE (37), p. 154.

COMPOSITION OF LITHOFELLIC ACID. a

	Ettling and Will.[*]			Wöhler.	Calculated $C_{40}H_{36}O_8$
Carbon	71·19	70·80	70·23	70·83	70·83
Hydrogen	10·85	10·78	10·95	10·60	10·48
Oxygen	17·96	18·42	18·92	18·57	18·69

a Ann. der Chem. und Pharm., XXXIX., 242, XLI., 154.

NOTE (38), p. 177.

COMPOSITION OF SOLANINE FROM THE BUDS OF GERMINATING POTATOES. *a*

	Blanchet.
Carbon	62·11
Hydrogen	8·92
Nitrogen	1·64
Oxygen	27·33

a Ann. der Pharm., VII., 150.

NOTE (39), p. 177.

COMPOSITION OF PICROTOXINE. *a*

	Francis.[*]
Carbon	60·26
Hydrogen	5·70
Nitrogen	1·30
Oxygen	32·74

a In another analysis, M. Francis obtained 0·75 per cent. of nitrogen. The picrotoxine employed for these analyses was partly obtained from the manufactory of M. Merck, in Darmstadt, and was partly prepared by M. Francis himself; it was perfectly white, and beautifully crystallized. Regnault, as is well known, found no nitrogen in this compound.

NOTE (40), p. 177.

COMPOSITION OF QUININE.

	Liebig.[*]	Calculated $C_{20}H_{12}NO_2$
Carbon	75·76	74·39
Hydrogen	7·52	7·25
Nitrogen	8·11	8·62
Oxygen	8·62	9·64

Y 2

NOTE (41), p. 177.

COMPOSITION OF MORPHIA. *a*

	Liebig.*	Regnault.		Calculated $C_{35}H_{20}NO_6$
Carbon	72·340	72·87	72·41	72·28
Hydrogen ...	6·366	6·86	6·84	6·74
Nitrogen......	4·995	5·01	5·01	4·80
Oxygen	16·299	15·26	15·74	16·18

a Ann. der Pharm., XXVI., 23.

NOTE (42), p. 177.

COMPOSITION OF CAFFEINE, THEINE, GUARANINE, THEOBROMINE, AND ASPARAGINE.

	Caffeine. *a* Pfaff and Liebig.*	Théine. *b* Jobst.	Guaranine. *c* Martius.	Calculated $C_8 H_5 N_2 O_2$
Carbon	49·77	50·101	49·679	49·798
Hydrogen ...	5·33	5·214	5·139	5·082
Nitrogen ...	28·78	29.009	29·180	28·832
Oxygen ...	16·12	15·676	16·002	16·288

a Ann. der Pharm., I., 17.
b   Do.   XXV., 63.
c   Do.   XXVI., 95.

Guaranine is the name given to the crystallized principle of the guarana officinalis, till it was shewn to be identical with caffeine and théine, as the above analyses demonstrate.

COMPOSITION OF THEOBROMINE. *a*

	Woskreseusky.			Calculated $C_9 H_5 N_3 O_2$
Carbon	47.21	46·97	46·71	46·43
Hydrogen ...	4·53	4·61	4·52	4·20
Nitrogen ...	35·38	35·38	35·38	35·85
Oxygen......	12·88	13·04	13·39	13·51

a Ann. der Chem. und Pharm., XLI., 125.

COMPOSITION OF ASPARAGINE. *a*

	Liebig.*	Calculated $C_8 H_8 N_2 O_6 + 2HO$
Carbon	32·351	32·35
Hydrogen	6·844	6·60
Nitrogen	18·734	18·73
Oxygen.........	42·021	42·32

a Ann. der Pharm., VII., 146.

ON THE CONVERSION OF BENZOIC ACID INTO HIPPURIC ACID.*

By Wilhelm Keller.

(From the Annalen der Chemie und Pharmacie.)

So early as in the edition of Berzelius's "Lehrbuch der Chemie," published in 1831, Professor Wöhler had expressed the opinion, that benzoic acid, during digestion, was probably converted into hippuric acid. This opinion was founded on an experiment which he had made on the passage of benzoic acid into the urine. He found in the urine of a dog which had eaten half a drachm of benzoic

---

* To the evidence produced by A. Ure, of the conversion of benzoic acid into hippuric acid in the human body, M. Keller has added some very decisive proofs, which I append to this work on account of their physiological importance. The experiments of M. Keller were made in the laboratory of Professor Wöhler, at Göttingen; and they place beyond all doubt the fact that a non-azotised substance taken in the food can take a share, by means of its elements, in the act of transformation of the animal tissues, and in the formation of a secretion. This fact throws a clear light on the mode of action of the greater number of remedies; and if the influence of caffeine on the formation of urea or uric acid should admit of being demonstrated in a similar way, we shall then possess the key to the action of quinine and of the other vegetable alkalies.—J. L.

acid with his food, an acid crystallizing in needle-shaped prisms, which had the general properties of benzoic acid, and which he then took for benzoic acid. (Tiedemann's Zeitschrift für Physiologie, i. 142.) These crystals were obviously hippuric acid, as plainly appears from the statements, that they had the aspect of nitre, and, when sublimed, left a residue of carbon. But at that time hippuric acid was not yet discovered; and it is well known, that till 1829, when these acids were first distinguished from each other by Liebig, it was uniformly confounded with benzoic acid.

The recently published statement of A. Ure, that he actually found hippuric acid in the urine of a patient who had taken benzoic acid, recalled this relation, so remarkable in a physiological point of view, and induced me to undertake the following experiments, which, at the suggestion of Professor Wöhler, I made on myself. The supposed conversion of benzoic acid into hippuric acid has, by these experiments, been unequivocally established.

I took, in the evening before bedtime, about thirty-two grains of pure benzoic acid in syrup. During the night I perspired strongly, which was probably an effect of the acid, as in general I am with great difficulty made to transpire profusely. I could perceive no other effect, even when, next day, I took the same dose three times; indeed, even the perspiration did not again occur.

The urine passed in the morning had an uncommonly strong acid reaction, even after it had been evaporated, and had stood for twelve hours. It deposited only the usual sediment of earthy salts. But when it was mixed with muriatic acid, and allowed to stand, there were formed in it long prismatic, brownish crystals, in great quantity, which, even in this state, could not be taken for benzoic acid. Another portion, evaporated to the consistence of syrup, formed, when mixed with muriatic acid, a magma of crystalline scales. The crystalline mass was pressed, dissolved in hot water, treated with animal charcoal, and recrystallized. By this means the acid was obtained in colourless prisms, an inch in length.

These crystals were pure hippuric acid. When heated, they melted easily; and when exposed to a still stronger heat, the mass was carbonised, with a smell of oil of bitter almonds, while benzoic acid sublimed. To remove all doubts, I determined the proportion of carbon in the crystals, which I found to be 60·4 per cent. Crystallized hippuric acid, according to the formula $C_{18}H_8NO_5 + HO$, contains 60·67 per cent. of carbon; crystallized benzoic acid, on the other hand, contains 69·10 per cent. of carbon.

As long as I continued to take benzoic acid, I was able easily to obtain hippuric acid in large quantity from the urine; and since the benzoic acid seems so devoid of any injurious effect on the health, it would be easy in this way to supply one's self with large quantities of hippuric acid. It would only be necessary to engage a person to continue for some weeks this new species of manufacture.

It was of importance to examine the urine which contained hippuric acid, in reference to the two normal chief constituents, urea and uric acid. Both were contained in it, and apparently in the same proportion as in the normal urine.

The inspissated urine, after the hippuric acid had been separated by muriatic acid, yielded, on the addition of nitric acic, a large quantity of nitrate of urea. It had previously deposited a powder, the solution of which in nitric acid gave, when evaporated to dryness, the well-known purple colour characteristic of uric acid. This observation is opposed to the statement of Ure; and he is certainly too hasty in recommending benzoic acid as a remedy for the gouty and calculous concretions of uric acid. He seems to suppose that the uric acid has been employed in the conversion of benzoic acid into hippuric acid; but as his observations were made on a gouty patient, it may be supposed that the urine, even without the internal use of benzoic acid, would have been found to contain no uric acid. Finally, it is clear that the hippuric acid existed in the urine in combination with a base, because it only separated after the addition of an acid.

INDEX.

INDEX.

A.

Acid.

—*Acetic.* Composition; and relation to that of aldehyde, 279, 280.

—*Benzoic.* Composition, and relation to that of oil of bitter almonds, 279, 280. Converted into hippuric acid in the human body, 150, 325.

—*Carbonic.* Is the form in which the inspired oxygen and the carbon of the food are given out, 13. Its formation in the body the chief source of animal heat, 17—22. Occurs combined with potash and soda, in the serum of the blood, 41. Formed by the action of oxygen on the products of the metamorphosis of the tissues, 60. Its formation may also be connected with the production of fat from starch, 85—91. Generated by putrefaction of food in the stomach of animals, 115. Also by the fermentation of bad wine in man, when it causes death by penetrating into the lungs, 116. Escapes through both skin and lungs, *ib.* Produced, along with urea, by the oxidation of uric acid, 140. Produced, with several other compounds, by the oxidation of blood, *ib.* May be formed, along with choleic acid, from hippuric acid, starch, and oxygen, 152. Also, along with choleic acid, urea, and ammonia, by the action of water and oxygen on starch and proteine, *ib.* Produced, along with fat and urea, from proteine, by the action of water and oxygen, in the absence of soda, 154. Combines with the compound of

ACID.

iron present in venous blood, and is given off when oxygen is absorbed, 269. Is absorbed by the serum of blood in all states, 270.

—*Cerebric*. Its composition, 184. Its properties, 186.

—*Choleic*. Represents the organic portion of the bile, 133. Its formula, 134. Its transformations, 125. Half its formula, added to that of urate of ammonia, is equal to the formula of blood $+$ a little oxygen and water, 136. Produced in the oxidation of blood, 140. Views which may be taken of its composition, 148. May be formed by the action of oxygen and water on proteine and starch, 152. Products of its oxidation, 154. Various ways in which it may be supposed to be formed in the body, 160. Its composition, 315. Cannot be said to exist ready formed in the bile, 317.

—*Cholic*. Its composition, 318. Derived from choleic acid, 134, 135. Possible relation to choleic acid, 148.

—*Choloidic*. Its composition, 315. Derived from choleic acid, 135. Possible relation to choleic acid, 148. Possible relation to starch, 157. Possible relation to proteine, 141.

—*Cyanic*. Its formula, 281.

—*Cyanuric*. Its formula, 281.

—*Hippuric*. Its composition, 319. Appears in the urine of stall-fed animals, 82. Is destroyed by exercise, 82, 139. Is probably formed in the oxidation of blood, 140. Is found in the human urine after benzoic acid has been administered, 150, 325. May be derived from proteine when acted on by oxygen and uric acid, 151. With starch and oxygen, it may produce choleic and carbonic acids, 152. May be derived from the oxidation of choleic acid, 154.

—*Hydrocyanic* or *Prussic*. Its poisonous action explained, 274.

—*Lithofellic*. Its composition, 322. Probably derived from the oxidation of choleic acid : is the chief constituent of bezoar stones, 154.

—*Lactic*. Its composition, 309. Its origin, 111. Does not exist in the healthy gastric juice, 112.

—*Margaric*. Exists in bile, 317.

—*Muriatic*. Exists in the free state in the gastric juice, 109, 112. Is derived from common salt, 112, 161.

—*Oxaluric*. Analysis of, 321.

Acid.

—*Parabanic.* Analysis of, 321.

—*Phosphoric.* Exists in the urine of the carnivora in considerable quantity, 78, 163. Its proportion very small in that of the graminivora, 79. Derived from the phosphorus of the tissues, 78. It is retained in the body to form bones and nervous matter, 80.

—*Sulphuric.* Exists in the urine of the carnivora, 78, 163. Derived from the sulphur of the tissues, 78.

—*Uric.* Its composition, 318. Products of its oxidation, alloxan, oxalic acid, carbonic acid, urea, &c., 137, 140. Is probably derived, along with choleic acid, by the action of oxygen and water on blood or muscle, 136. Disappears almost entirely in the system of man and of the higher animals, 55, 137. Appears as calculus, when there is a deficiency of oxygen, 137. Never occurs in pthisical cases, *ib.* Yields mulberry calculus when the quantity of oxygen is somewhat increased, but only urea and carbonic acid with a full supply of oxygen, *ib.* Uric acid calculus promoted by the use of fat and of certain wines, 139. Unknown on the Rhine, *ib.* Uric acid and urea, how related to allantoine, 141; to gelatine, 142. Forms the greater part of the urine of serpents, 54. Yields, with the elements of proteine and oxygen, hippuric acid and urea, 151. How related to taurine, 155, 156. Calculi of it never occur in wild carnivora, but often in men who use little animal food, 146.

Affinity, Chemical. Is the ultimate cause of the vital phenomena, 9, 10. Is active only in the case of contact, and depends much on the order in which the particles are arranged, 205. Its equilibrium renders a compound liable to transformations, 207. In producing the vital phenomena, it is modified by other forces, 209. It is not alone the vital force or vitality, but is exerted in subordination to that force, 232.

Air. Introduced into the stomach during digestion with the saliva, 113. Effects of its temperature and density, dryness, &c., in respiration, 15, 16.

Albumen. Animal and vegetable albumen identical, 47, 48. Their composition, 293, 294, 308, 309. Vegetable albumen, how obtained, 45. Is a compound of proteine, and in organic composition identical with fibrine and caseine, 47, 104, 106.

Exists in the yolk as well as the white of eggs, 107. Also in the serum of the blood, 41. Is the true starting point of all the animal tissues, 107, 108.

ALCOHOL. Is hurtful to carnivorous savages, 179. Its mode of action : checks the change of matter, 239. In cold climates serves as an element of respiration, 22.

ALDEHYDE. Its composition ; how related to that of acetic acid, 279, 280.

ALKALIES. Mineral alkalies essential both to vegetable and animal life, 164. Vegetable alkalies all contain nitrogen, all act on the nervous system, and are all poisonous in a moderate dose, 177, 182. Theory of their action : they take a share in the transformation or production of nervous matter, for which they are adapted by their composition, 182—189. Action of caustic alkalies on bile, or choleic acid, 134.

ALLANTOINE. Is found in the urine of the fœtal calf. How derived from proteine. How related to uric acid and urea, 141. How related to choleic acid, 148. Its composition, 319.

ALLEN and PEPYS. Their calculation of the amount of inspired oxygen, 283.

ALLOXAN. Formed by the oxidation of uric acid, 137. Converted by oxidation into oxalic acid and urea, oxaluric and parabanic acids, or carbonic acid and urea, ib. How related to taurine, 156. Seems to act as a diuretic. Recommended for experiment in hepatic diseases, ib. (note).

ALMONDS, BITTER. Oil of. Its composition ; how related to benzoic acid, 280.

AMMONIA. Combined with uric acid it forms the urine of serpents, birds, &c., 54. Its relation to choleic, choloidic, and cholic acids, 135. Is one of the products which may be formed by the oxidation of blood, 140; or of proteine, 152. Its relation to uric acid, urea, and taurine, 155. To allantoine and taurine, 155, 156. To alloxan and taurine, 156. To choleic and choloidic acid and taurine, 158. To urea, water, and carbonic acid, 159. Is found in combination with acids in the urine of the carnivora, 163.

ANALYSIS. Of dry blood, 283, 314. Of dried flesh, 314. Of fæces, 285. Of black bread, ib. Of potatoes, ib. Of peas, ib. Of beans, ib. Of lentils, ib. Of fresh meat, ib. Of moist bread, ib. Of moist potatoes, ib. Of the fibrine and albumen of blood, 293, 310, 311. Of vegetable fibrine and albumen, vegetable caseine and gluten, 294, 295. Of animal caseine, 295. Of starch, 296, 297. Of grape or starch sugar, 297. Of sugar of milk, 298. Of gum, *ib.* Of oats, *ib.* Of hay, 299. Of fat, 300. Of cane-sugar, *ib.* Of cholesterine, *ib.* Of wax, 307. Of cyanic acid, cyanuric acid, and cyamelide, 308. Of aldehyde, metaldehyde, and elaldehyde, 307. Of proteine, 308. Of albumen from the yolk and white of egg, *ib.* Of lactic acid, 309. Of gas from the stomach of cows after eating to excess, *ib.* Of gas from stomach and intestines of executed criminals, *ib.* Of gelatinous tissues, 311. Of tissues containing chondrine, 312. Of arterial membrane, *ib.* Of horny tissues, *ib.* Of the lining membrane of the egg, 313. Of feathers, *ib.* Of the pigmentum nigrum, *ib.* Of choleic acid, 315. Of taurine, *ib.* Of choloidic acid, *ib.* Of cholic acid, 318. Of uric acid, *ib.* Of alloxan, 319. Of urea, *ib.* Of hippuric acid, *ib.* Of allantoine, *ib.* Of xanthic oxide, 320. Of cystic oxide, *ib.* Of oxalic acid, 320. Of oxaluric acid, *ib.* Of parabanic acid, *ib.* Of roasted flesh, 322. Of lithofellic acid, *ib.* Of solanine, 323. Of picrotoxine, *ib.* Of quinine, *ib.* Of morphia, 324. Of caffeine, theine, or guaranine, *ib.* Of theobromine, *ib.* Of asparagine, 325.

ANIMAL HEAT. Derived from the combination of oxygen with the carbon and hydrogen of the metamorphosed tissues, which proceed ultimately from the food, 17, 18. Is highest in those animals whose respiration is most active, 19. Is the same in man in all climates, 19, 20. Is kept up by the food in proportion to amount of external cooling, 22. Is not produced either by any direct influence of the nerves, or by muscular contractions, 29—34. Its amount in man, 34. Chemical action the sole source of it, 38. The formation of fat from starch or sugar must produce heat, 91, 94. The elements of the bile, by combining with oxygen, serve chiefly to produce it, 61.

ANIMAL LIFE. Distinguished from vegetable life by the absorption of oxygen, and the production of carbonic acid, 2. Must not be confounded with consciousness, 6, 7. Conditions necessary to animal life, 9, 12. Depends on an equilibrium between waste and supply, 245, 254, 265.

ANTISEPTICS. They act by putting a stop to fermentation, putre-

59, 61. The process of assimilation in adult and young carnivora compared, 67. Their urine, 78. The assimilative process in adult carnivora less energetic than in graminivora, 80.
They are destitute of fat, 82. They swallow less air with
their food than graminivora, 118. Concretions of uric acid
are never found in them, 146. Both soda and ammonia found
in their urine, 163.

CASEINE. One of the azotised nutritious products of vegetable
life, 47. Abundant in leguminous plants, 47. Identical in
organic composition with fibrine and albumen, 47, 48. Animal
caseine found in milk and cheese : identical with vegetable
caseine, 51. Furnishes blood to the young animal, 52. Is
one of the plastic elements of nutrition, 96. Yields proteine,
105, 106. Its relation to proteine, 126. It contains sulphur,
ib. Potash essential to its production, 164. Contains more
of the earth of bones than blood does, 52. Its analysis, 295.

CEREBRIC ACID. See ACID, *Cerebric.*

CHANGE OF MATTER. See METAMORPHOSIS OF TISSUES.

CHEMICAL ATTRACTION. See AFFINITY.

CHEVREUL. His researches on fat, 84. His analysis of fat, 300 ;
of cholesterine, *ib.*

CHLORIDE OF SODIUM. See COMMON SALT.

CHOLEIC ACID. See ACID, *Choleic.*

CHOLESTERINE. See BILE.

CHOLIC ACID. See ACID, *Cholic.*

CHOLOIDIC ACID. See ACID, *Choloidic.*

CHONDRINE. Its relation to proteine, 126. Analysis of tissues
containing it, 312.

CHRONIC DISEASES. The action of inspired oxygen is the cause
of death in them, 27, 28.

CHYLE. When it has reached the thoracic duct, it is alkaline,
and contains albumen coagulable by heat, 145.

CHYME. It is formed independently of the vital force, by a chemical transformation, 108. The substance which causes this
transformation is derived from the living membrane of the
stomach, 109. Chyme is acid, 145.

CLOTHING. Warm clothing is a substitute for food to a certain
extent, 22. Want of clothing accelerates the rate of cooling,
and the respirations, and thus increases the appetite, *ib.*

COLD. Increases the appetite by accelerating the respiration, 22.

metamorphosis derived from the transformation of a substance proceeding from the lining membrane of the stomach, 109. The oxygen introduced with the saliva assists in the process, 113. Lactic acid has no share in it, 111, 112.

DISEASE. Theory of, 254 *et seq.* Cause of death in chronic disease, 27. Disease of liver caused by excess of carbon or deficiency of oxygen, 23. Prevails in hot weather, 24.

DOG. Amount of bile secreted by, 64. Digests the gelatine of bones, 97. His excrements contain only bone earth, 98. Concretion of urate of ammonia said to have been found by Lassaigne in a dog, doubtful, 146 (*note*).

DUMAS. His analysis of choleic acid, 315; of choloidic acid, *ib.*; of taurine, *ib.*; of cholic acid, 318; of hippuric acid, 319.

E.

EGGS. Albumen of the white and of the yolk identical, 107. Analysis of both, 308; of lining membrane, 313. The fat of the yolk may contribute to the formation of nervous matter, 108. This fat contains iron, 107.

ELALDEHYDE. See ALDEHYDE.

ELEMENTS. Of nutrition, 96. Of respiration, *ib.*

EMPYREUMATICS. They check transformations, 170. Their action on ulcers, 121.

EQUILIBRIUM. Between waste and supply of matter is the abstract state of health, 245, 258. Transformations occur in compounds in which the chemical forces are in unstable equilibrium, 109.

ETTLING. His analysis of wax, 307. ETTLING and WILL. Their analysis of lithofellic acid, 322.

EXCREMENTS. Contain little or no bile in man and in the herbivora, none at all in the dog and other carnivora, 64. Those of the dog are phosphate of lime, 98. Those of serpents are urate of ammonia, 54. Those of birds also contain that salt, 54. Those of the horse and cow compared with their food, 290, 291.

EXCRETIONS. Contain, with the secretions, the elements of the blood or of the tissues, 132—136. Those of the horse and cow compared with their food, 290, 291. Bile is not an excretion, 63.

F.

FÆCES. Analysis of, 285.

FAT. Theory of its production from starch, when oxygen is deficient, 83 *et seq.*; from other substances, 86. The formation of fat supplies a new source of oxygen, 89; and produces heat, 90 *et seq.* Maximum of fat, how obtained, 94. Carnivora have no fat, 82. Fat in stall-fed animals, 89. Occurs in some diseases in the blood, 95. Fat in the women of the East, 99. Composition compared with that of sugar, 84, 85. Analysis of fat, 300. Disappears in starvation, 25. Is an element of respiration, 96.

FATTENING OF ANIMALS. See FAT.

FEATHERWHITE WINE. Its poisonous action, 116.

FEBRILE PAROXYSM. Definition of, 256.

FEHLING. His analysis of metaldehyde and elaldehyde, 307.

FERMENTATION. May be produced by any azotised matter in a state of decomposition, 120. Is arrested by empyreumatics, *ib.* Is analogous to digestion, 119.

FEVER. Theory and definition of, 256.

FIBRE. Muscular. See Flesh.

FIBRINE. Is an element of nutrition, 96. Animal and vegetable fibrine are identical, 45. Is a compound of proteine, 105. Its relation to proteine, 126. Convertible into albumen, 42. Is derived from albumen during incubation, 107. Its analysis, 293, 294, 311. Vegetable fibrine, how obtained, 45, 46.

FISHES. Yield phosphuretted hydrogen, 191 (*note*).

FLESH. Consists chiefly of fibrine, but, from the mixture of fat and membrane, has the same formula as blood, 133. Analysis of flesh, 314, 322. Amount of carbon in flesh compared with that of starch, 77, 299.

FOOD. Must contain both elements of nutrition and elements of respiration, 96. Nutritious food, strictly speaking, is that alone which is capable of forming blood, 40. Whether derived from animals or from vegetables, nutritious food contains proteine, 44, 106 *et seq.* Changes which the food undergoes in the organism of the carnivora, 53 *et seq.* The food of the herbivora always contains starch, sugar, &c., 70. Food, how dissolved, 108 *et seq.* Azotised food has no direct influence on the formation of uric acid calculus, 138. Effects of super-

pound of proteine, and cannot form blood, 127 *et seq.* May serve as food for the gelatinous tissues, and thus spare the stomach of convalescents, 98, 130. In starvation the gelatinous tissues remain intact, 97. Its relation to proteine, 126. Its formula, 142. Its analysis, 311, 322.

GOEBEL. His analysis of gum, 298.

GLOBULES of the blood are the carriers of oxygen to all parts of the body, 171—175. They contain iron, 265 *et seq.*

GLUTEN. Contains vegetable fibrine, 46. Analysis of it, 295.

GMELIN. On the sugar of bile, 147.

GOOSE. How fattened to the utmost, 94.

GRAMINIVORA. See HERBIVORA.

GRAPE-SUGAR. An element of respiration, 96. Is identical with starch sugar and diabetic sugar, 72. Its composition, 73. Its analysis, 297.

GROWTH, or increase of mass, greater in graminivora than in carnivora, 80. Depends on the blood, 40; and on compounds of proteine, 106. See NUTRITION.

GUM. An element of respiration, 96. Its composition, 73. Is related to sugar of milk, *ib.* Its analysis, 298.

GUNDLACH. His researches on the formation of wax from honey by the bee, 301.

H.

HAIR. Analysis of, 312. Its relation to proteine, 126. Analysis of proteine from hair, 308.

HAY. Analysis of, 299.

HEPATIC DISEASES. Cause of, 23.

HERBIVORA. Their blood derived from compounds of proteine in their food, 48. But they require also for their support non-azotised substances, 70. These last assist in the formation of their bile, 147 *et seq.* They retain the phosphoric acid of their food to form bone and nervous matter, 80. Their urine contains very little phosphoric acid, 79. The energy of vegetative life in them is very great, 81. They become fat when stall-fed, 82.

HESS. His analysis of wax, 307.

HYBERNATING ANIMALS. Their fat disappears during the winter sleep, 25. They secrete bile and urine during the same period, 61.

of proteine alone are nutritious, 106. Occurs when the vital
force is more powerful than the opposing chemical forces, 198.
Theory of it, 210. Is almost unlimited in plants from the absence of nerves, 212. Depends on the momentum of force
in each part, 227. Depends also on heat, 243.

O.

OATS. Amount required to keep a horse in good condition, 74.
Analysis of, 298.

OIL OF BITTER ALMONDS. Its composition. How related to
benzoic acid, 279, 280.

OLD AGE. Characteristics of, 248 *et seq*.

OPPERMANN. His analysis of wax, 307.

ORGANS. The food of animals always consists of parts of organs,
2. All organs in the body contain nitrogen, 42, 43. There
must exist organs for the production of nervous matter, 189;
and the vegetable alkalies may be viewed as food for these
organs, *ib*.

ORGANISED TISSUES. All contain nitrogen, 42, 43. All such as
are destined for effecting the change of matter are full of small
vessels, 223. Their composition, 126. The gelatinous and
cellular tissues, and the uterus, not being destined for that
purpose, are differently constructed, 224. Waste of organised
tissues rapid in carnivora, 76.

ORIGIN. Of animal heat, 17, 31. Of fat, 81 *et seq*. Of the
nitrogen exhaled from the lungs, 114 *et seq*. Of gelatine, 127
et seq., 143. Of uric acid and urea, 135 *et seq*. Of bile, 135,
143, 146 *et seq*., 159. Of hippuric acid, 150, 325. Of the
chief secretions and excretions, 152. Of the soda of the bile,
161 *et seq*. Of the nitrogen in bile, 168. Of nervous matter,
183 *et seq*.

ORTIGOSA. His analysis of starch, 297.

OXALIC ACID. A product, along with urea, of the partial oxidation of uric acid, occurring in the form of mulberry calculus,
137. Its analysis, 321.

OXYGEN. Amount consumed by man daily, 12, 283. Amount
consumed daily in oxidising carbon by the horse and cow, 14.
The absorption of oxygen characterizes animal life, 2. The
action of oxygen is the cause of death in starvation and in
chronic diseases, 25—28. The amount of oxygen inspired

varies with the temperature, dryness, and density of the air,
16. Is carried by arterial blood to all parts of the body, 171.
Fat differs from sugar and starch only in the amount of oxygen, 84. It also contains less oxygen than albumen, fibrine,
&c., 86. The formation of fat depends on a deficiency of oxygen, 88 *et seq.*; and helps to supply this deficiency, 89. Oxygen essential to digestion, 113. Relation of oxygen to some
of the tissues formed from proteine, 126. Oxygen and water,
added to blood or to flesh, yield the elements of bile and of
urine, 135. Action of oxygen on uric acid, 136, 139 ; on
hippuric acid, 82, 139 ; on blood, 140 ; on proteine, with uric
acid, 151 ; on proteine and starch, with water, 152 ; on choleic acid, 154 ; on proteine, with water, 154. By depriving
starch of oxygen and water, choloidic acid may be formed, 157.
Oxygen is essential to the change of matter, 173. Its action
on the azotised constituents of plants when separated, 213.
Its action on the muscular fibre essential to the production of
force, 220—226. Oxygen is absorbed by hybernating animals,
241. Is the cause of the waste of matter, 243 ; and of animal
heat, 244, 252. Blood-letting acts by diminishing the amount
of oxygen which acts on the body, 258. Its absorption is the
cause of the change of colour from venous to arterial blood,
265. The globules probably contain oxide of iron, protoxide in venous blood, peroxide in arterial, 267 *et seq*. All
parts of the arterial blood contain oxygen, 173, 174, 266,
271.

P.

PEARS. Analysis of starch from unripe, 297.

PEAS. Form part of the diet of soldiers in Germany, 287, 289.
Abound in vegetable caseine, 47. Analysis of peas, 285 ; of
starch from peas, 296.

PEPYS and ALLEN. Their calculation of the amount of inspired
oxygen, 283.

PEROXIDE OF IRON. Probably exists in arterial blood, 267 *et seq.*

PFLÜGER. His analysis of the gas obtained by puncture from the
abdomen of cattle after excess in green food, 309.

PHENOMENA of motion in the animal body, 195 *et seq.*

PHOSPHATES. See BONES.

PHOSPHORIC ACID. See ACID, *Phosphoric.*

PHOSPHORUS. Exists in albumen and fibrine, 41, 48, 126. It is

Q.

R.

S.

T.

U.

Printed by J. L. Cox & Sons, 75, Great Queen Street.

Printed in the United States
By Bookmasters